THE INVISIBLE CREATION

THE INVISIBLE CREATION

LOOKING INTO THE DEEP

SHUVADIP GANGULI

THE INVISIBLE CREATION
BY
SHUVADIP GANGULI

First published in India by Notionpress
Interior and cover designed by Shuvadip Ganguli

This book is dedicated
To
All physics lovers with or without a
physics background

"Man must rise above the Earth—to the top of the atmosphere and beyond—for only thus will he fully understand the world in which he lives."

— Socrates, Philosopher

CONTENTS

PREFACE

I am always interested in writing books. Books in which I can put my scientific thoughts that keep my mind buzzing. I try to keep my books as lucid as possible to all my readers. My book 'PHYSICS My love: Story of Physics for Everyone' was my first attempt. It was an elementary introduction to the fascinating world of Physics. It received lots of love and acceptance from you all readers, which made me more enthusiastic to write the second one. In this book, I try to unwrap the hidden mysteries of our Universe, which is, I think almost an irresistible attraction for every science student.

Did you know that the stars, planets, galaxies and other visible objects, taken all of them together make up only about four percent of our Universe? The other ninety-six percent is made of invisible stuff. Stuff we cannot see or detect them but now we know that they are there, hidden, deep inside somewhere but they are present everywhere. It's just a matter of fact that they are untraceable by our strongest and most sensitive devices.

'Black hole', 'Dark Matter', 'Dark Energy', 'Worm Hole', 'Multiverse', 'time travel' are some of the most popular terms which almost everyone has come across. It is our human instinct, to be more curious to know about what we cannot see or feel. Nowadays we might also see the above-mentioned mysterious things in many science-fiction stories and movies. But, if we come to know the real explanation behind those things in the language physics, then we can enjoy them more.

Why this book?

When I was a school student, I read many books based on space and cosmology, but none of them could cover the entire realm of it. As I grew older a passion grew inside me – Why not write a book which will provide knowledge to all the basic facts encompassing this field of Physics. The book would be a step by step measure to gather all information and provide all of it to the reader rather than to buy different books.

For example, if you read a book about the Black hole, then you may not get a sufficient idea about the theory of relativity in that book or vice versa. But, relativity is one of the most important pillar theories to understand the concept of the black hole. Now if you want to understand the General theory of relativity you have to understand 'gravity' properly. But for gravity, you have to search for another book. Many textbooks of astrophysics cover them all, but with complex mathematics which are unpleasant for general readers or beginners.

In this book, you will get all the topics stitched together in one place. I have tried my best to provide all of them in one platter starting from the tiniest to the biggest.

What is inside?

The title of my book might give you a bit of idea that in this book we will focus on the deepest invisible mysteries of the Universe.

In the first chapter, you will get a brief concept of light and its properties. Then in the next chapter, the Relativity theory will make you more curious. In this chapter, I will mainly talk about the Special Theory of Relativity. Chapter 3 is based on the concept of space and time. Before the General Theory of Relativity, you should learn about Gravity. So, I put those two in chapter 4. Chapters 5 and 6 are about Black hole and the method used to obtain the first image of black hole. The next chapter is about white hole and worm hole. In chapter 8, I have discussed one of the most mysterious objects called Dark matter. But, before I discuss dark energy which is another most mysterious object, I decided to explain our much-known Doppler Effect, redshift and important Hubble's law in chapter 9 and chapter 10 respectively. Then you'll know about Cosmological constant and most mysterious Dark energy in chapter 11. In the last chapter we will see our Universe from the beginning to the end again and at last, will get an idea about its destiny. This book has two parts: first is the main chapter and the second part is the appendix. In the appendix part, I included some important theories and their simple mathematical derivations. Those who are interested in the mathematical derivations can read this part.

When I decided to write this book there was thought always in my mind, to explain everything in an easy way for all types of readers (not only for science students or teachers). The physics of Black hole, White hole, Dark matter, and Dark energy is very complex and hard to understand for general readers. But, in this book, I try to represent those as simple as possible with various examples and illustrations without many mathematical derivations. I hope, this book may be of interest to students, teachers, as well as general readers who are interested in astronomy. All valuable suggestions for the improvement of the book will be highly appreciated.

January 2, 2020   Shuvadip Ganguli

FOREWORD

Shuvadip's first book "PHYSICS My Love" initiated us to the fascinating world of Physics. It was a good read and hooked its readers expecting his next book. The second book "The Invisible Creation" is a continuum of his earlier treatise. While the former dealt with the evaluation of classical physics to quantum physics and some other basic information about physics, this book deals with the deeper and broader intricacies of the nature of light, relativity, gravitational wave, Black hole etc. Very much like the first book, the present book is also characterised by its remarkable lucidity of language and logic of principles and practice and tone. He discusses at length the evaluation of the idea of 'Black hole' and also lays bare the two-century-old controversy as to the duel character of light.

The book is liberally sprinkled with interesting pictures and mathematical deductions so as to make reading more fun.

What heartens me more is his approach to the genre of popular science writing. It is lucid and hardly intimidating. It draws readers to the inner intricacies of Physics.

Kishor Kumar Chowdhury
Asst. Master in Physics
Bankura zilla School
W.B.; India

ACKNOWLEDGEMENT

I would like to express my deep and sincere gratitude to my most respected, dearest teacher and guide Mr. Kishor Kumar Chowdhury for giving me an interest in Physics at my childhood and providing priceless guidance. He has written the foreword of this book at my request.

I am especially thankful to one of my closest friends Md. Nazrul Islam (Physics teacher, Harrow Hall School, Park Street Kolkata-16). In spite of his busy schedule, he helped me in editing the manuscript, proofing of the book and made this book more interesting for readers.

I am very grateful for encouraging response from the numerous readers who are interested in this field.

I am extremely grateful to my parents for their love, prayers, caring and sacrifices for educating and preparing me for the future. Also, I express my thanks to my grandfather, who always encourage me to write books. They all kept me going, and this book would not have been possible without them.

CHAPTER 1
NATURE OF LIGHT

"Light brings us the news of the Universe"

Sir William Bragg

The main cause of the invisibility (or darkness) is the lack of light. Unless light from an object reaches our eye, the object cannot be seen. Light is an external physical agency that produces the sensation of sight. Light itself, however, is invisible!

Light is an electromagnetic wave. Electromagnetic radiation is transmitted in waves or particles at different wavelengths and frequencies. This broad range of wavelengths is known as the electromagnetic spectrum. Our eyes are sensitive to a narrow band of the electromagnetic spectrum, known as the visible light spectrum. Visible light is a form of electromagnetic (EM) radiation in a particular frequency (or wavelength) range.

White is the result of a mixture of two or more colours of light. We see that when a white light passing through a prism; it divided into 7 colours (VIBGYOR as in rainbow).

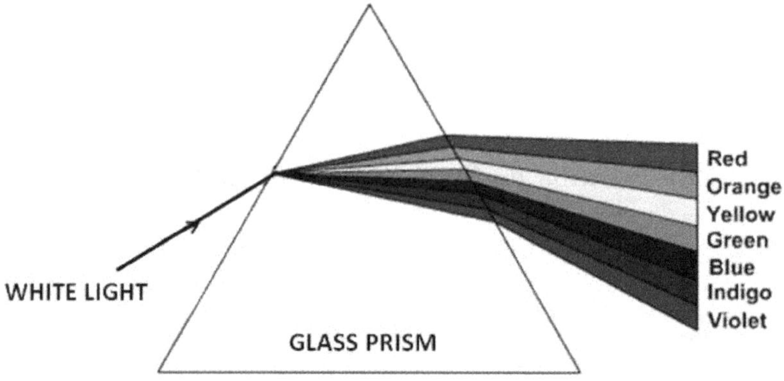

Figure 1.1

Black is the absence of the visible light spectrum wavelengths. The human eye can only see visible light, but the light comes in many other colours like radio, infrared, ultraviolet, X-ray, and gamma-ray. These are invisible to the naked eye.

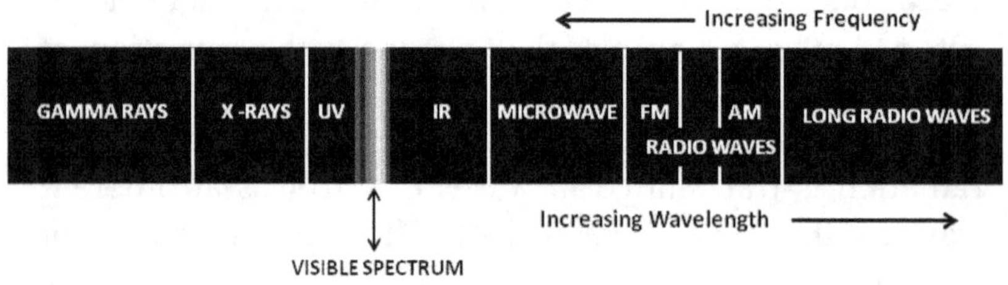

Figure 1.2

Astronomers can determine what objects are made of because each element absorbs light at specific wavelengths, called an absorption spectrum. By knowing the absorption spectra of elements, astronomers can use spectroscopes to determine the chemical composition of stars, dust clouds and other distant objects. But unfortunately, in the case of black hole it doesn't work because no light can be escape from a black hole. Here we need some advanced mechanism. We'll come to this point later. Now, we will concentrate to the nature of the light and the theories developed according to its nature.

A hypothetical medium once believed to be necessary to support the propagation of electromagnetic radiation, called Ether. It is now regarded as unnecessary. The existence of the ether was first called into question as a result of the Michelson-Morley experiment (discussed later).

The first theories of light were developed nearly simultaneously. They are Huygens' wave theory and Newton's corpuscular theory.

According to Newton, the sensation of sight was due to the impact on retina of swarms of minute particles emitted by the source of light. These particles travelled in a straight line in homogeneous media and were either attracted or repelled when they approached the surface bounding two media. In 1704, he proposed the particle theory of light in his book 'Opticks'.

Newton's theory could not explain interference, diffraction and polarization. To double refraction, the corpuscular theory had no answer. This had been explained by Huyghens.

Christiaan Huyghens (Dutch mathematician, physicist, and astronomer) opposed the theory of Newton and gave us the wave theory of light in 1670. According to him, light is a wave or waveform. Huyghens had a very important insight into the nature of wave propagation which is nowadays called Huygens' Principle. This principle states that,

Every point of a wave-front may be considered as the source of small secondary wavelets, which spread out in all directions from their centres with a velocity equal to the velocity of propagation of the wave. The new wave-front is the surface that touches the secondary wavelets.

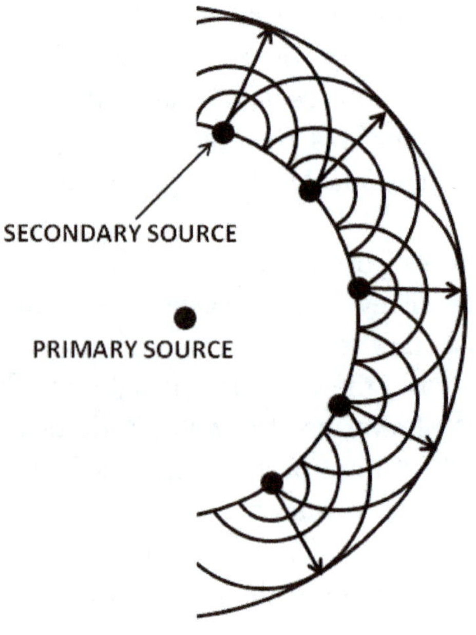

Figure 1.3

Augustin Fresnel (French civil engineer and physicist) extended Huyghens' Principle and explained on the wave theory the phenomena of interference, diffraction, polarization and double refraction.

Maxwell postulated his electromagnetic theory of light in 1867. The experimental evidence in support of this theory was obtained by Hertz in 1888. Maxwell found that these electromagnetic waves could have any wavelength and would travel in a vacuum with the speed of light. It was a transverse wave. From these, he concluded that light waves must be electromagnetic in nature.

The idea of the quantum theory of light arises from Max Planck's work on the explanation of the distribution of the energy in the spectrum of a black-body (1900). He made some assumptions which are called Planck's hypothesis or Planck's postulates. Planck made the assumption that energy was made of individual units or quanta. He assumed that emission of light energy by a source was a discontinuous process, small but definite amount being emitted in at a time. But, he did not suggest how this light energy distributed. In 1905, Albert Einstein published a paper entitled, "On a heuristic point of view concerning the production and transformation of light" in the 'Annalen der Physik',(Page:132–148). The paper proposed light with particle-like properties and explains the recently discovered photoelectric effect. This led to the quantum revolution in physics and earned Einstein the Nobel Prize in Physics in 1921. Einstein took the clue from Planck's explanation of the blackbody spectrum and proposed a simple explanation of this effect. He assumed that the energy of light is not distributed over the whole expanding wavefront but rather is concentrated into small regions. Each of these 'packages' has an energy $h\nu$ (ν is the frequency of light and h is the Planck's constant). In this model, the light energy comes in pieces, each of the amount hv. Each of these 'pieces' represents the smallest quantity of energy or the quantum of light of that frequency. But Einstein did not give the name Photon. In 1926, the optical physicist Frithiof Wolfers and the chemist Gilbert N. Lewiscoined gave the name photon for these particles.

Now just think about the point, what light really is? Is it a particle or a wave?

The wave theory successfully explains the propagation, interference, diffraction and polarization of light. But when we look to phenomena which involve the interaction of light with matter, wave theory does not work. Some of the more well known of these phenomena are the origin of spectra, Photoelectric Effect, Compton Effect etc. The concept of the photon has successfully explained all such phenomena. So, we can see that both the wave model and the particle model have definite limitations. Each of them helps us to visualize a different aspect of the phenomenon of light. By accepting them both we can explain with perfect satisfaction the observed facts. So, the conclusion is light exhibits a dual nature.

CHAPTER 2
THEORY OF RELETIVITY

"Einstein, in the special theory of relativity, proved that different observers, in different states of motion, see different realities."

Leonard Susskind

Albert Einstein one of the greatest minds of the 20th century and his relativity theory is one the most significant theory in the history of science. It changed the landscape of science by introducing revolutionary concepts that shook our understanding of the physical world. Relativity means the quality or state of being relative. The same incident can be seen differently by different observers. As for example, imagine you're standing on a train while your friend is standing outside the train, watching it pass by. If lightning struck on both ends of the train, your friend would see both bolts of lightning strike at the same time. But on the train, you are closer to the bolt of lightning that the train is moving toward. So you see this lightning first because the light has a shorter distance to travel. This thought experiment showed that time moves differently for someone moving than for someone standing still.

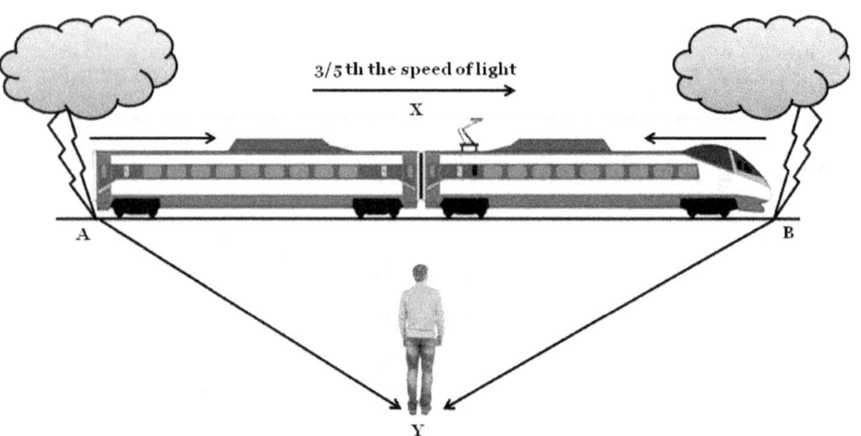

Figure 2.1: The light from A will arrive at X before that from B. Hence X will observe the lightning at A as happening before that at B. Y, however, will observe both lightning to be simultaneous.

Do you know that the concept of relativity was not introduced by Einstein? It was first introduced by an Italian astronomer, physicist and engineer Galileo Galilei. Einstein's major contribution was the recognition that the speed of light in a vacuum is constant. This idea does not have a major impact on our lives since we travel at speeds much slower than light speed. But, for objects travelling near light speed, Newtonian and Galilean relativity does not hold. At that point, we have to apply Albert Einstein's theory of relativity.

Before Einstein, physicists understood the universe in terms of three laws of motion presented by Isaac Newton in 1687 (in his famous book *Philosophiæ Naturalis Principia Mathematica*, Latin for Mathematical Principles of Natural Philosophy). We all know these famous three laws of motion from high school.

In Newtonian Mechanics, there is no upper limit for the speed of any particle. There comes a lot of problems when two frames are moving with respect to each other at a considerable speed of light. The addition of velocities that is Galilean Transformation shows a big problem where some particles can even travel faster than light and thus everything seems against basic laws of nature. So, we are in need of a different type of mechanics.

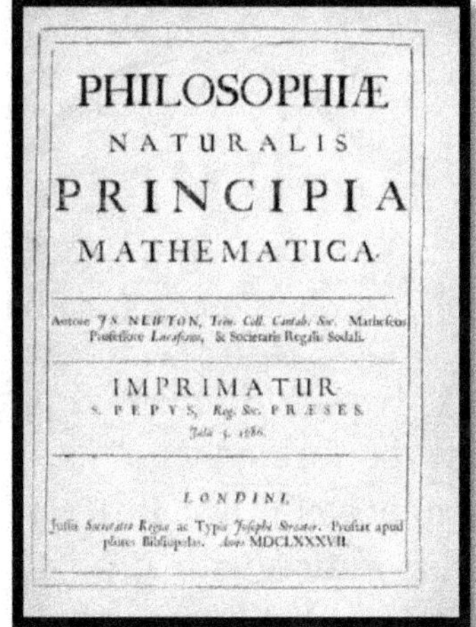

Figure 2.2: Sir Isaac Newton and Title page of 'principia', first edition (1687)

In the 19th century, scientists imagined that space was filled everywhere by a continuous medium called the ether. Light rays and radio signals were waves in this ether. They thought light travel at a fixed speed through the ether. That means its speed would appear to be lower if you were travelling in the same direction as the light and its speed would appear to be higher if you were travelling in the opposite direction to the light. But a series of experiments failed to find any evidence for differences in speed due to motion through the ether.

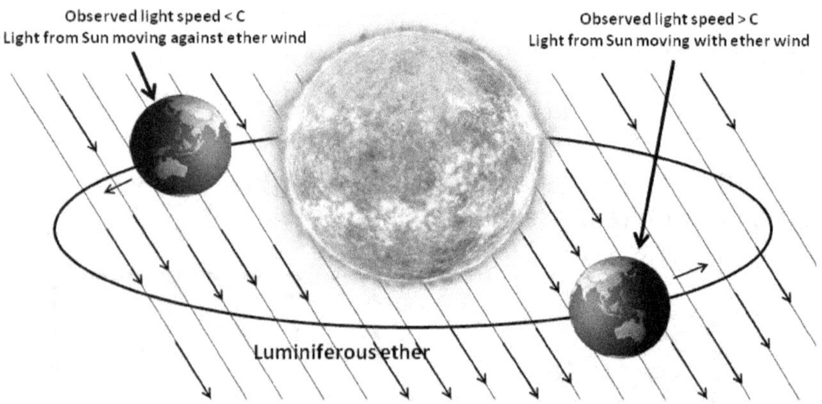

Figure 2.3: Earth moving around Sun in the ether medium

A.A.Michelson invented optical interferometer. Michelson first performed the experiment in 1881 to show the existence of Ether. In 1887 he performed the experiment in collaboration with E.W.Morley. They computed the phase difference between the beams 1 and 2 (see figure 1.4).

This difference can arise due to two causes: a. the different path lengths travelled and b. the different speeds of travel with respect to the instrument because of the Ether wind. The second cause is important for this moment.

In Michelson-Morley experiment* they mounted the interferometer on a massive stone slab for stability and floated the apparatus in mercury (Hg) so that it could be rotated smoothly about a central pin. To make the light path as long as possible, mirrors were arranged on the slab to reflect the beams back and forth through eight round trips. Observations were made day and night and during all seasons of the year. But, the expected fringe shift was not observed. The experimental conclusion was that there was no fringe shift at all. This proves the non-existence of Ether. There is no Ether wind!

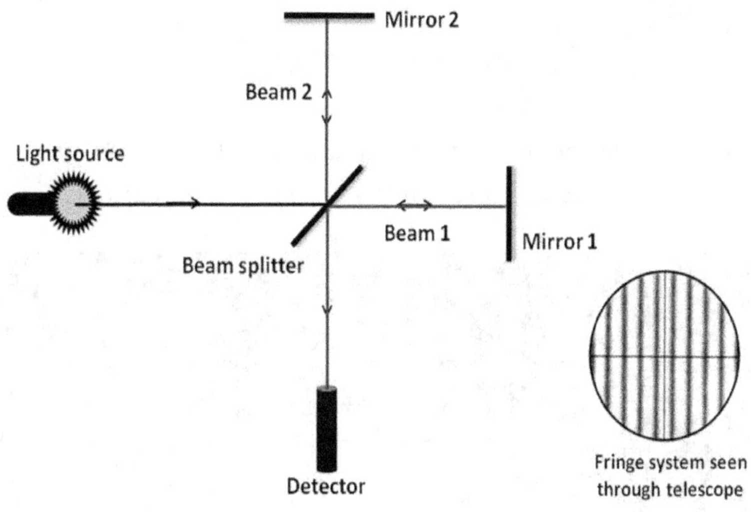

Figure 2.4: Michelson- Morley experiment diagram

---

* For detailed calculation of Michelson-Morley experiment see Appendix-1 (Page: 147-150)

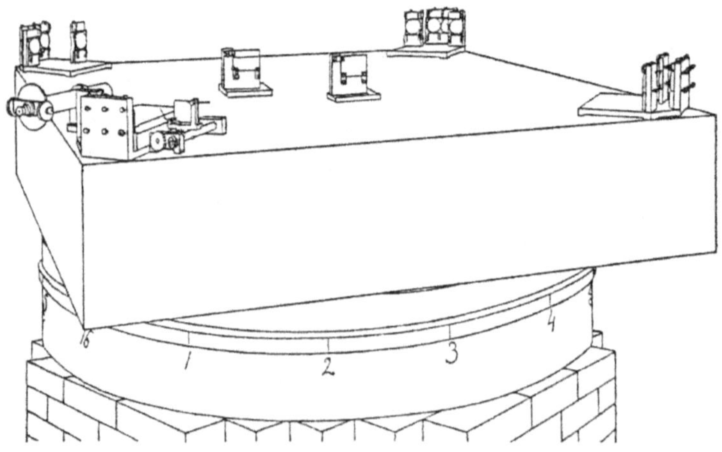

Figure 2.5: Michelson and Morley experiment setup, mounted on a stone slab and floating in a pool of mercury

Figure 2.6: The setup of Michelson Morley experiment in 1887 at what is now Case Western Reserve University
Credit: Case Western Reserve University/ Wikimedia Commons

This experiment showed that the light velocity is the same when measured along two perpendicular axes in the frame which, presumably, is moving relative to ether-frame in a different direction at different times of the year.

George FitzGerald and the Hendrik Lorentz were the first to suggest that bodies moving through the ether would contract and that clocks would slow. This shrinking and slowing would be such that everyone would measure the same speed for light no matter how they were moving with respect to the ether.

It was a clerk of the Swiss Patent Office in Bern, named Albert Einstein, who removed the idea of ether and solved the previous problems about the speed of light. Einstein began to think about how moving observers see events differently from stationary observers. In my book *'PHYSICS My love: Story of Physics for Everyone'*, I mentioned about some of Einstein's thought experiments and most groundbreaking discoveries *(chapter 6)*.

Einstein said that 'absolute motion' is a meaningless concept. He published his theory, special relativity, in the German physics journal *'Annalen der Physik'* in 1905.

The special theory of relativity is based on two postulates:

1. *The principle of relativity*: The laws of physics are the same in all inertial reference frames. It says that the laws of physics are absolute, universal, and the same for all inertial observers.

2. *The principle of the constancy of the speed of light*: The speed of light in free space has the same value c in all inertial reference frames.

Why the special relativity called 'special'? Special Theory of Relativity is called 'Special' because it is limited to a special case of inertial frames of reference. Einstein's postulates require a new consideration of the fundamental nature of time and space. In this section, we discuss how the postulates affect measurements of time and length intervals by observers in different frames of reference.

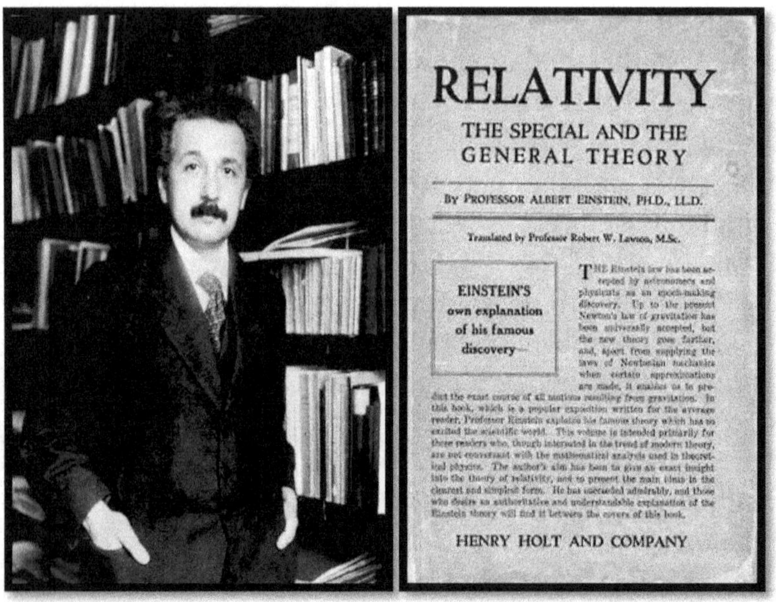

Figure 2.7: Albert Einstein (left) and the original 1920 English publication of his paper (right)

But, at first, we have to know about 'frame of reference'. Generally, a place from where observations are made is called frame of reference. In physics, a frame of reference is represented by a co-ordinate system with respect to which co-ordinates of any point in space is determined and is also associated with a clock to measure any event in space. There are two types of frame of reference. One is inertial and another is non-inertial. Two frames are said to be inertial if they are at rest or moving with constant velocity with respect to each other. A frame of reference is said to be non-inertial if it is either accelerating or rotating. Here I said two inertial frames because single absolute inertial frame doesn't exist. Whereas single or absolute non-inertial frame exists. You can get many examples of different frames in any standard textbooks.

Now let us suppose that an event occurred at a point M (x,y,z,t) in an inertial frame of reference S. In another inertial frame S', the same event will be specified by four other quantities x', y', z' and t'.

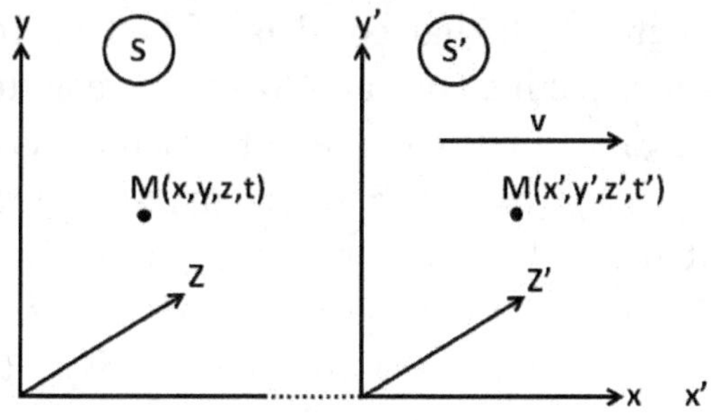

Figure 2.8

Now the S' frame moves along the x-axis with a velocity v with respect to S frame. Since each set of four quantities in S' frame corresponds to the set of four quantities in S frame, the two sets must be related by a relationship. These equations are called the *Lorentz transformation*.

$$x' = \frac{x - vt}{\sqrt{1 - \frac{v^2}{c^2}}}$$

$$y' = y$$

$$z' = z$$

$$t' = \frac{t - \frac{vx}{c^2}}{\sqrt{1 - \frac{v^2}{c^2}}}$$

The fact that the speed of light is the same in all reference frames has the consequence that moving clocks run slow. It is called *'time dilation'*.

Imagine a light clock that consists of two mirrors and a beam of light reflecting back and forth between the mirrors (see figure 2.9). At first, you are in the same inertial frame as the light clock. You are therefore measuring the proper time, denoted Δt_0. The Mirrors are separated by distance 'd'. The light moves with a speed 'c'. One "tick" is when the light goes from one mirror to the other and back again. The proper time for one tick is given by

$$\Delta t_0 = \frac{2d}{c}$$

You are in a different inertial frame to the light clock. The light clock is moving with velocity u ms⁻¹ in the x-direction. The path the light takes is now different as it has only a finite speed. The time taken for one tick is denoted Δt, the observed time.

$$\Delta t = \frac{\Delta t_0}{\sqrt{1 - \frac{u^2}{c^2}}}$$

Now we'll solve an example to understand this effect properly. Muons are elementary particles with a (proper) lifetime of 2.2 µs. They are produced with very high speeds in the upper atmosphere when cosmic rays (high energy particles from space) collide with air molecules. Let the height L_0 of the atmosphere to be 100 km in the reference frame of the Earth.

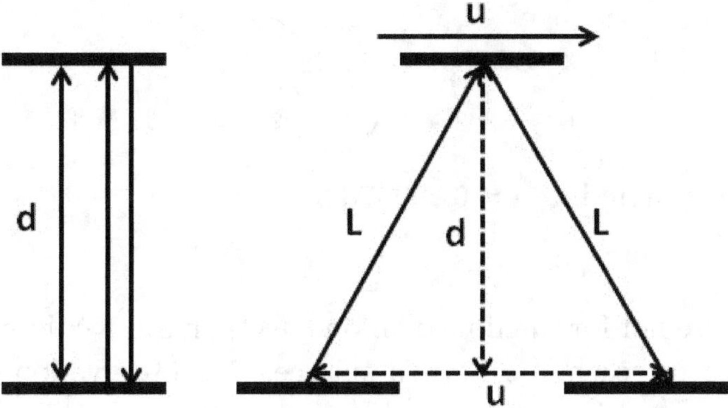

Figure 2.9: One "tick" is when the light goes from one mirror to the other and back again (left)
The light clock is moving with velocity u ms⁻¹ in the x-direction (right)

Now we have to find the minimum speed that enables the muons to survive the journey to the surface of the Earth.
The birth and decay of the muon can be considered as the 'ticks' of a clock. In the frame of reference of the Earth (observer O) this clock is moving, and therefore its ticks are slowed by the time dilation effect. If the muon is moving at a speed that is close to c, the time necessary for it to travel from the top of the atmosphere to the surface of the Earth is

$$\Delta t = \frac{L_0}{C} = 333\mu s$$

If the muon is to be observed at the surface of the Earth, it must live for at least 333 µs in the Earth's frame of reference. In the muon's frame of reference, the interval between its birth and decay is a proper time interval of 2.2 µs. Now,

$$333\mu s = \frac{2.2\mu s}{\sqrt{1 - \frac{u^2}{c^2}}}$$

Solving, we find $\quad u = 0.999978\,c$

If it were not for the time dilation effect, muons would not survive to reach the Earth's surface. The observation of these muons is a direct verification of the time dilation effect of special relativity.

Another important consequence is that a moving object's length is measured to be shorter than its proper length, which is the length as measured in the object's own rest frame. This phenomenon is called 'length contraction'. By similar calculation as time dilation, we can obtain the expression for length contraction.

$$\Delta l = \Delta l_0 \sqrt{1 - \frac{u^2}{c^2}}$$

See figure 2.9 to understand the effect of length contraction in a better way. This affects every moving object like a running bus, train, even when you're running. But, under normal circumstances, the effect is so incredibly tiny that it's not noticeable at all. An object needs to be moving at speeds approaching the speed of light before length contraction starts to take a great affect on it.

There is another type of time dilation, called 'gravitational time dilation' which is observed near black holes. But before we discuss it, we have to know about the general theory of relativity.

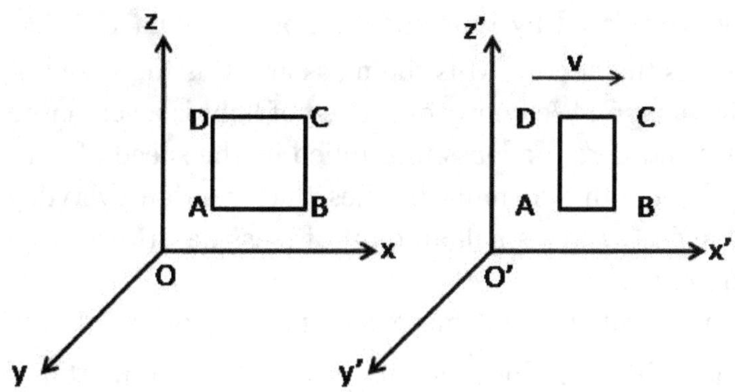

Figure 2.10: In the above figure (a), ABCD is a square at rest according to observer O. But if it moves with some velocity along X-axis, the length of the arm AB and DC will contract due to length contraction. Now the shape of the square as observed by observer O' like figure (b).

Relativity theory is a very complex and important part of physics. It cannot be compressed in one chapter. There are so many topics and complex mathematical calculations in it.

But, I think, you are able to understand the next chapters if you understand this chapter properly. This is the basic relativity in a nutshell.

There is one more thing, which is very important, as it changed the whole concept of mass and energy. Almost everyone knows about the *'Mass-Energy Equivalence'*. Well, you don't know what this is? I think you know the famous equation $E=MC^2$.

According to this theory, anything having mass has an equivalent amount of energy and vice versa. Mass and energy are related by Einstein's famous formula $E=MC^2$, where E is the energy, M is the mass and C is the speed of light in vacuum. We know that speed of light is a very large number and here the mass multiplied by the speed of light squared. So, this formula implies that even an everyday object at rest with a small amount of mass has a very large amount of energy.

Einstein formulated this famous equation (not exactly the formula $E=MC^2$) in his relativity paper 'Does the Inertia of an object Depend Upon Its Energy Content?' in 1905.

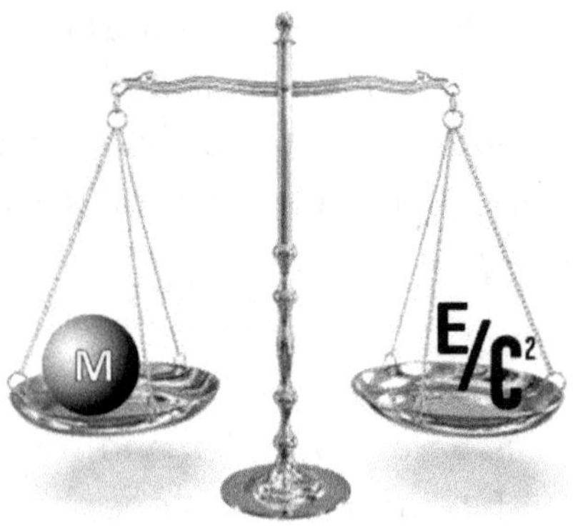

We all know about the Doppler Effect for sound waves from our school level. Light waves or any other electromagnetic waves also show Doppler Effect (Since light also emanates in wavelengths, this means that the wavelengths can stretch or crunch together depending on the relative position of objects. That said, we don't notice it on the daily-life-sized scale because light travels so much faster than the speed of sound — a million times faster.) Redshift and Blueshift are two very important terms in the case of the light wave. There are two more types of this effect called Relativistic redshift and Gravitational redshift. These two are very important concepts that we need for further chapters in this book. But, this part will be discussed in appendix-2 (Page: 151-154) in detail.

CHAPTER 3
SPACE AND TIME
OR
SPACE-TIME?

"Time and space are not conditions of existence,
time and space is a model of thinking."
Albert Einstein

In classical physics the time coordinate is left unaffected upon transformation from one inertial frame of reference to another. In relativistic mechanics, the time coordinate in one inertial frame depends on the time and the space coordinates of another inertial frame (as in Lorentz transformation). Therefore, in relativity, the time t and space coordinates x, y, z are treated together. This is known as the space-time geometry of four-dimensional space.

It was first proposed by the mathematician Hermann Minkowski in 1908 as a way to reformulate Albert Einstein's special theory of relativity (1905). By combining space and time into a single manifold we get a four-dimensional continuum called Minkowski space. On 5 November 1907 (a little more than a year before his death), Minkowski introduced his geometric interpretation of space-time in a lecture to the Göttingen Mathematical society with the title, 'The Relativity Principle'. On 21 September 1908, Minkowski presented his famous talk, 'Space and Time' to the German Society of Scientists and Physicians. 'Minkowski Diagram' is a method of visualizing spacetime. The defining feature of a Minkowski diagram is that light rays are drawn at a 45-degree angle to the line or plane representing space. The two most common types of Minkowski Diagrams are *'Lineland Minkowski Diagrams'* and *'Flatland Minkowski Diagrams'*. To characterize any phenomenon at a space position and at a certain time, we need to know the spatial and time coordinates in a four-dimensional space-time, as one usually refers to the set of all possible events. In such space-time, an event is a point with coordinates (x, y, z, t) or as most conventionally used (x, y, z, ct), the c is the absolute value of light speed in free space. Minkowski space-time is a graphical representation of events and sequences of events in space-time as seen by an observer at rest. Such sequences are named *'world lines'*.

This shows the space-time diagram in two dimensions (x, ct). World line OW represents a light ray travelling along the x-axis towards positive values of x (notice that OW bisects the quadrant formed by ct-axis and x-axis); GO represents a light ray travelling along the x-axis towards negative values of x. World lines for light are light lines. World line OP represents anything travelling with moderate constant speed compared to light speed (the slope gives a measure of the speed); OQ represents anything (except light) travelling with constant speed close to that of light; IJ represents events happening simultaneously; The ct-axis itself represents anything standing still at point O (x,ct) = (0,ct); The x-axis represents events happening at (x,ct) = (x,0).

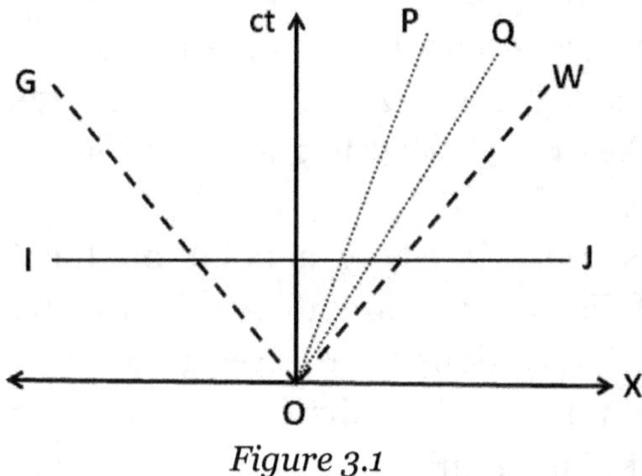

Figure 3.1

Now see figure 2.2. It represents the space-time diagram in three dimensions (x,y,ct). The light lines form a light cone. The circles cantered at ct-axis define simultaneity planes. The four dimensions case is impossible to be represented on a piece of paper. However, most of the interesting effects between events in four-dimensional space-time are reasonably represented in two and three-dimensional space-times.

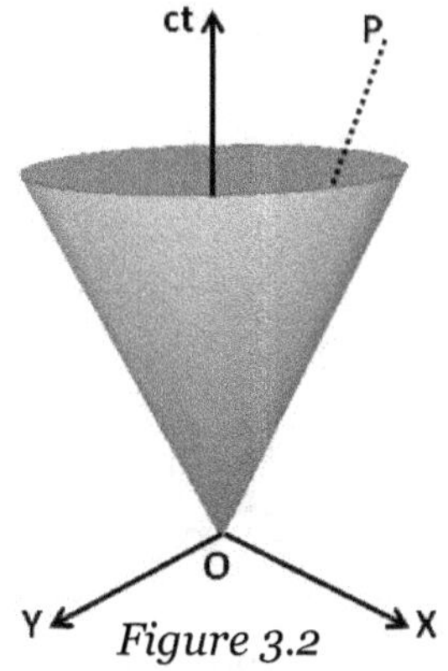

Figure 3.2

The light cone divides the four-dimensional space into two invariant separate domains characterized by two inequalities :

$$S^2 = c^2t^2 - x^2 - y^2 - z^2 < 0$$
$$S^2 = c^2t^2 - x^2 - y^2 - z^2 > 0$$

Where 'S' is called four-dimensional interval between the origin and the point (x, y, z, ct). We call this interval space-like, time-like or null according as S^2 is less than, greater than or zero.

An event is represented by a point on the Minkowski diagram. If light cones are drawn in the positive and negative time directions from a certain event (E), space-time is separated into three distinct regions: 'future', 'past', and 'present'. The future is the locus of all events that have not yet happened that E can (and does) affect. The reason that E cannot affect anything outside of its future light cone is because, in order to do so, it would have to send some sort of message to the desired location faster than the speed of light. This is impossible, for the speed of light is believed to be the greatest possible speed. The past is the locus of events that contributed to E state. Anything that happened before E and is not in the past light cone could not have affected E.

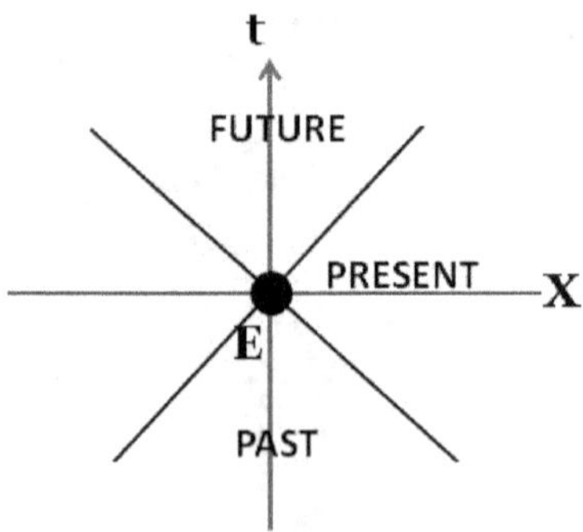

Figure 3.3: Space-time diagram in two dimensions

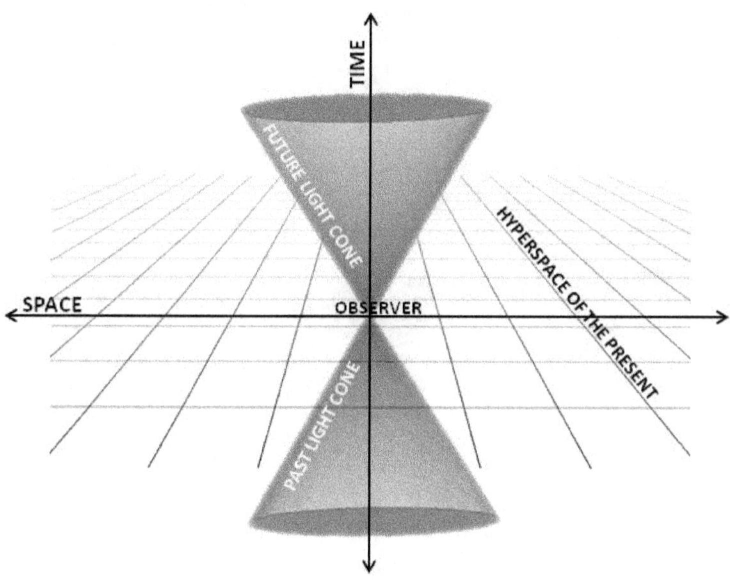

Figure 3.4: Space-time diagram in three dimensions with the time coordinate. The four dimensions case is impossible to be represented on a piece of paper.

CHAPTER 4
GRAVITY

"Gravitation cannot be held responsible for people falling in love"

Albert Einstein

In physics, there are four types of fundamental interactions, also known as fundamental forces. These are 'Gravitational', 'Electromagnetic', 'Strong' and 'Weak'. In this chapter, our main focus is on gravitational interaction or force. It is the weakest of all the fundamental interactions but has infinite range. Electromagnetic interaction has also infinite range as gravitational but cancels at large distances as very large objects have no charges, they are uncharged. So only gravity has an infinite range and it is attractive for all types of matter.

Galileo Galilei first studied gravity scientifically. Galileo discovered through this experiment that ignoring air resistance, the objects fell with the same rate, proving his prediction true, while at the same time disproving Aristotle's theory of gravity, which states that objects fall at speed proportional to their mass.

According to Newton, the force of gravity is always attractive, works instantaneously at a distance, and has an infinite range. Most importantly, it affects everything with mass - and has nothing to do with an object's charge or chemical composition. It acts between all bodies having mass, following the long-range inverse square type Newtonian law of gravitation.

$$F = \frac{Gm_1m_2}{r^2}$$

Where m_1 and m_2 are two masses separated by a distance r and G is the gravitational constant.

Newton said that gravity is a force due to mass and only mass responds to gravity. So, according to this theory light should not be affected by gravity as light is massless (since light is made of photons that have zero mass). But in the next part of this chapter, we'll see that it's not true. The second problem with Newton's theory was that it described gravity as an instantaneous force of attraction between two massive objects. Consequently, if you move one of them, the other knows about the movement immediately due to the change in gravitation, irrespective of the distance between them. So, if our Sun vanished, Earth will stop feeling its attraction immediately, which is also not true. Actually, Newton did not explain the mechanism of gravity properly. But, over the years, scientists in just about every discipline have tested Newton's laws of motion and found them to be amazingly predictive and reliable.

After about 200 years, Albert Einstein said that gravity is not a force but is the result of curvature of space-time. This statement had changed science's perception of gravity. Ten years later of the special theory of relativity, in 1915, Einstein included the effect of gravity to generalise this theory for all state of motion (with or without force). Because special relativity applies only when everything moves with a constant velocity (there are no forces or accelerations). Einstein became a legend after this discovery.

EFFECT OF GRAVITY ON SPACE-TIME AND LIGHT

"Mass tells space how to curve and space-time tells mass how to move."

John Wheeler

Light is a transverse, electromagnetic wave and does not have any mass. Light travels in a straight line. We know that Gravity affects objects which have some mass but light particles do not have any mass so gravity cannot affect it directly. But from relativity theory, we know about the space-time continuum, which is affected by gravity. Greater is the mass of an object, stronger is it's gravitational field and greater is the gravitational field, greater is the space-time curved around it.

Let's took a simple example to understand it in a better way. Imagine a sheet of rubber stretched out. Imagine that you put a heavy ball in the centre of the sheet. The weight of the ball will bend the surface of the sheet close to it. This is a two-dimensional picture of what gravity does to space-time in four dimensions (see figure 4.1). Now send a little marble rolling from one side of the rubber sheet to the other. Instead of the marble taking a straight path, it will follow the contour of the sheet that is curved by the weight of the ball in the centre. This is similar to how the gravitational field created by an object (the ball) affects light (the marble).

So even though light travels in a straight line, the straight path on which it was travelling is not straight anymore. The path is curved near massive objects which not the bending of light but the warping of space-time itself.

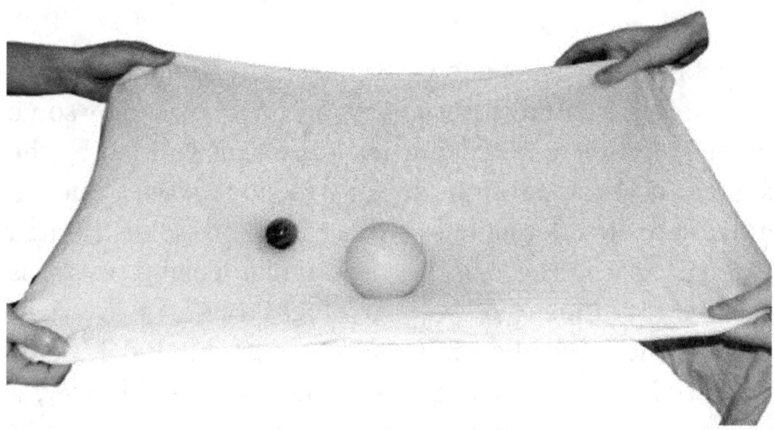

Figure 4.1: The weight of the heavy ball will bend the surface of the stretched rubber sheet. a little marble follows the contour of the sheet that is curved by the weight of the ball in the centre.

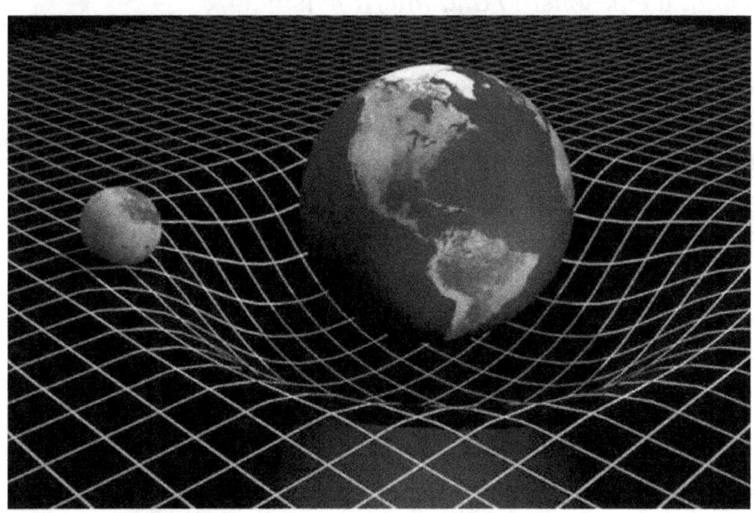

Figure 4.2: Earth (the heavy ball) and moon (the little marble) in space-time continuum.

GRAVITATIONAL LENSING

Gravitational lensing allows for the observation of a source of light that is not directly observable (it is blocked by some other object or material). Instead, the light emitted by the source is bent around an obstructing object due to gravitational interactions with other mass distributed through space. The reason gravitational lensing occurs is that light is influenced by gravity. Light travels in a straight path, which can be influenced by gravitational fields produced by other masses along its path. In this way, a source of light that is not typically visible along a straight line path can be seen through a curved path. It follows Albert Einstein's general theory of relativity.

There are three classes of gravitational lensing: *strong - lensing, weak-lensing and micro-lensing.*

Strong-lensing is the most visually stunning form of gravitational lensing. It requires an extremely massive object like Galaxy clusters. Partial arcs, full arcs (Einstein rings), and multiple images are all strong gravitational lensing features one can observe.

Weak gravitational lensing occurs much more frequently than strong lensing. Lenses can be clusters (in their outer regions), individual galaxies, or even large scale structure in the universe.

Micro-lensing is most common on the scale of the Milky Way galaxy. This can occur when background stars pass behind foreground stars.

This phenomenon also happened around a supermassive black hole, the arcs and rings around the black hole are the light from distant galaxies being distorted by the gravity of the black hole (see figure 4.5). The amount of gravitational lensing is directly related to the foreground object's mass. Gravitational lensing was first observed in 1979, but Einstein had suspected its possibility in 1912.

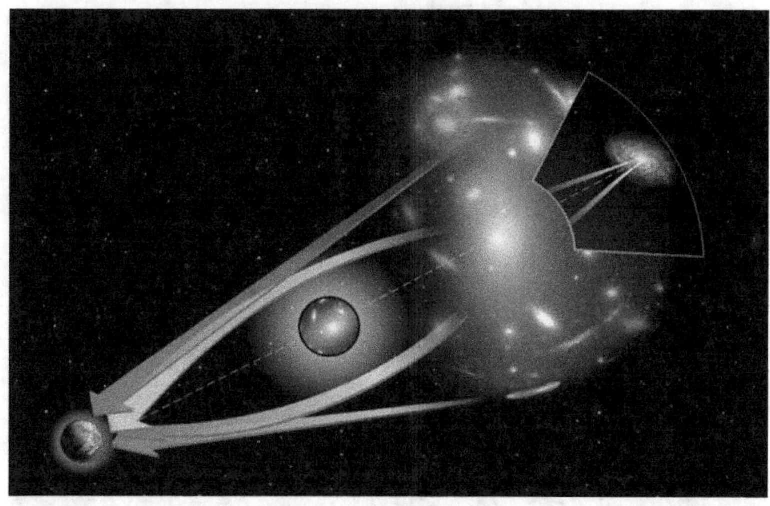

Figure 4.4: Illustration for the gravitational lens. Bending of light around a massive object from a distant source. The orange arrows show the apparent position of the background source. The white arrows show the path of the light from the true position of the source. (Picture courtesy: STScI)

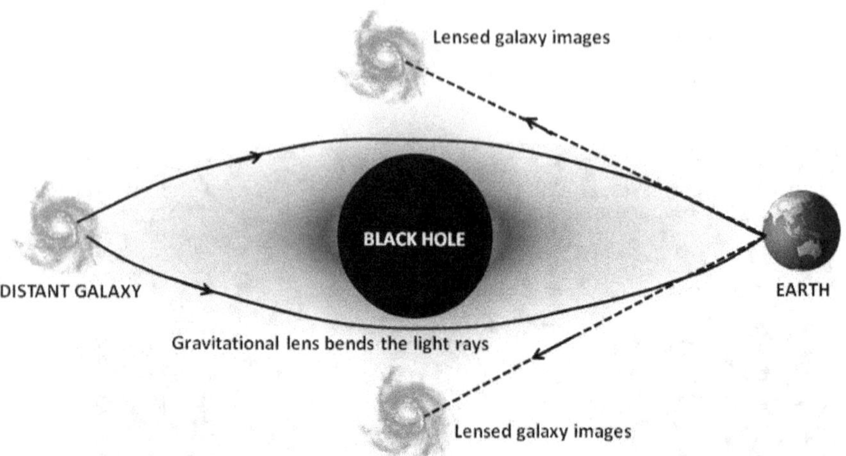

Figure 4.5

Figure 4.6: The gravity of a luminous red galaxy (LRG) has gravitationally distorted the light from a much more distant blue galaxy. (Picture courtesy: NASA)

GRAVITATIONAL WAVES

A binary black hole is a system consisting of two black holes in close orbit around each other. It may form during galaxy mergers. When these two black holes collide, it creates ripples in space-time known as gravitational waves. Gravitational waves are a fundamental prediction of the general theory of relativity. It is just like ripples on water surface in a pond (a small wave or series of waves on the surface of the water, especially as caused by an object dropping into it). See figure 4.7 and figure 4.8.

It is interesting that On 14 September 2015, the universe's gravitational waves were observed for the very first time. The waves, which were predicted by Albert Einstein a hundred years ago, came from a collision between two black holes. It took 1.3 billion years for the waves to arrive at the LIGO detector in the USA. LIGO, the Laser Interferometer Gravitational Wave Observatory, is a collaborative project with over one thousand researchers from more than twenty countries. The Nobel Prize committee has awarded the 2017 physics prize to three leaders of that project; Kip S. Thorne, Barry Barish, Rainer Weiss.

Figure 4.7: Water ripple in a pond

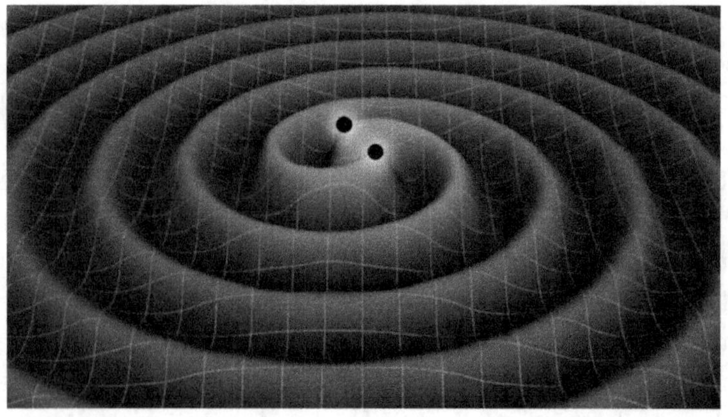

Figure 4.8: Two colliding black holes send ripples through the space-time fabric of the universe that are called 'gravitational waves'. (Picture courtesy: Hubblesite)

CHAPTER 5
BLACK HOLE

"Black holes are where God divided by zero"
Albert Einstein

Since primary science class, we are more inquisitive about the existence of something that we can't see. In astronomy, a Black hole is such an object that grows our interest. We all have assumed it as an invisible giant that sucks everything near it, from which nothing can escape – not even light. Yes, it is true, but partially. Let's define the term 'Black hole' in terms of physics. It is called 'black' because it absorbs all the light that enters it, there are no reflections just like a perfect black body in thermodynamics. In previous chapters, we learned about gravity and its effect on space-time. We know that, if the gravitational field increases, the space-time curvature also increases. So, space-time curved around massive objects because it has a larger gravitational field.

Figure 'a' shows the curvature of space-time due to the gravitational field of Earth. In figure 'b', a larger curvature of space-time arises due to more massive mass (more gravitational field) of the Sun. Figure 'c' shows the space-time curvature due to a black hole, where the gravitational field tends to infinity (see figure 5.1)

Black holes are like the wells with enormous gravitational pull. Those are incredibly massive and extremely dense but cover only a small region. Because of the relationship between mass and gravity, this means they have an extremely powerful gravitational force. This force is such powerful that even light can't escape from this region. Simply, a Black hole is a region of space-time with such a strong gravitational field that the escape velocity exceeds the speed of light (In physics, escape velocity is the minimum speed needed for a free object to escape from the gravitational influence of a massive body). It is a prediction of the General Theory of Relativity, which says that a sufficiently compact mass can deform space-time to form a black hole.

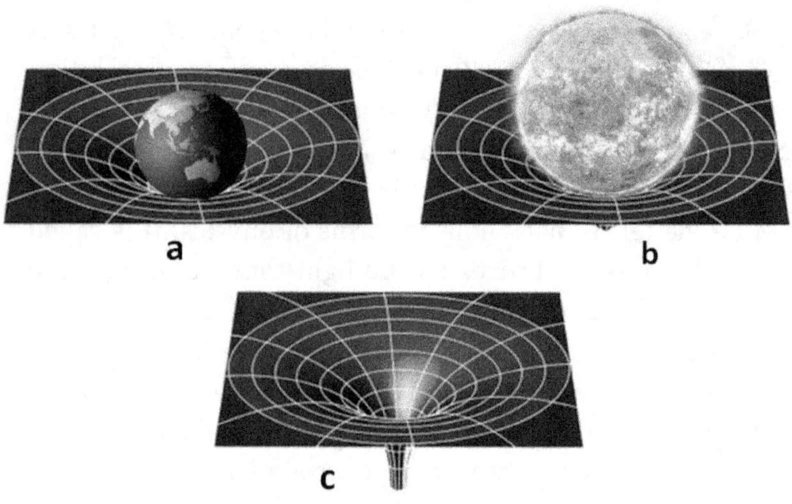

Figure 5.1

The boundary of the region from which no escape is possible is called the event horizon. Simply, it is like a shell of "points of no return", i.e., the boundary at which the gravitational pull of a massive object becomes so great as to make escape impossible. It is the boundary within which the black hole's escape velocity is greater than the speed of light. Everything that enters the event horizon falls farther into the hole and squashed into the singularity. It is also called the 'Schwarzschild barrier'. We cannot see the events that occur beyond this barrier or horizon. Here I like to mention an interesting thought. Just imagine that we send a person into the black hole and told him to send a message through radio signal about what he is seeing there. He also sends a message to us after just entering the event horizon. What do you think? Can we receive that message? The answer is NO!

Why? Because this signal is also an electromagnetic wave similar to the light and we know that escape velocity exceeds the speed of light beyond the event horizon. So, that signal must have a speed greater than the light to reach us.

But the description of event horizons given by general relativity is thought to be incomplete, because, event horizons are expected to have properties that are different from those predicted using general relativity alone. For this, we need both relativity and quantum mechanics (or Relativistic quantum mechanics*).

---

*In physics, relativistic quantum mechanics is any Poincaré covariant formulation of quantum mechanics. This theory is applicable to massive particles propagating at all velocities up to those comparable to the speed of light C and can accommodate massless particles.

STRUCTURE OF BLACK HOLE

For a non-rotating black hole, the radius of the event horizon is known as the Schwarzschild radius (By comparison, the Schwarzschild radius of the Earth is about the size of a marble). Any mass can be compressed sufficiently to form a black hole. The only requirement is that its physical size is less than the Schwarzschild radius. Our Sun would become a black hole if its mass was contained within a sphere about 3 km across.

All matter in a black hole is squeezed into a region of infinitely small volume, called the central singularity. A singularity is a point-like feature in space-time where the gravitational field becomes infinite. Everything within the event horizon is irreversibly drawn towards this point where the curvature of space-time becomes infinite and gravity is infinitely strong.

Some black holes spin around an axis and these are more complicated. It creates a cosmic whirlpool by dragging the whole around it and the event horizon is composed of two, instead of one, imaginary spheres. And there is a region called the 'ergo sphere' (*figure 5.2*).

Material, such as gas, dust or other stellar materials that have come close to a black hole but not quite fallen into it, forms a band of spinning matter around the event horizon called the 'accretion disk' (*figure 5.3*).

For some black holes magnetic field of high intensity is emitted perpendicular to the accretion disk creates jets of gas.

Albert Einstein first predicted black holes in 1916 with his general theory of relativity. The term "black hole" was given in 1967 by American astronomer John Wheeler, and the first one was discovered in 1971. It was the binary star Cygnus X-1.

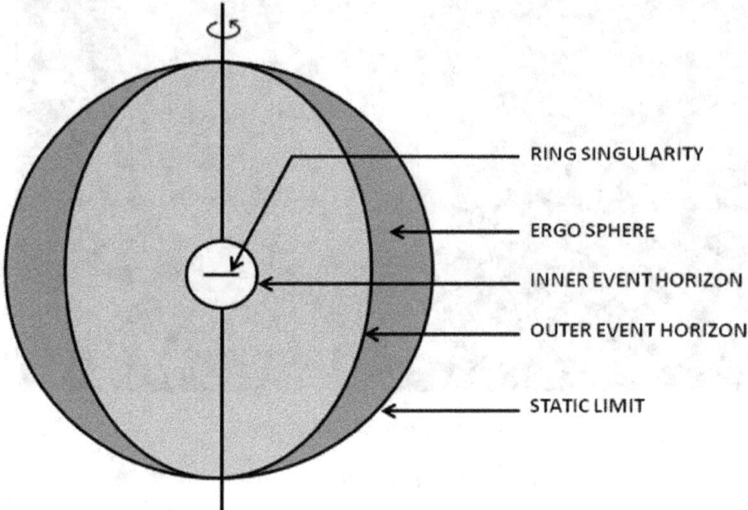

Figure 5.2: Structure of Black hole

John Wheeler

John Archibald Wheeler was an American theoretical physicist born on July 9, 1911. He was involved in the theoretical development of the atomic bomb.

He also contributed on Unified field theory, the space-time continuum, and gravitation. He first coined the term 'black hole'. For his work on nuclear fission and the technology of plutonium production, he was given the Fermi Award by the U.S. Atomic Energy Commission in 1968. He was awarded the Niels Bohr International Gold Medal in 1982.

Figure 5.3

Karl Schwarzschild

Karl Schwarzschild was a German astronomer, born on October 9, 1873. Schwarzschild first published his paper on the theory of celestial orbits at the age of 16 only. He gave many contributions in the development of 20th-century astronomy.

Schwarzschild gave the first exact solution of Albert Einstein's field equations. In 1916, he derived a quantity called 'Schwarzschild radius'. It is the radius of the event horizon of a black hole. Any object with a physical radius smaller than its Schwarzschild radius will be a black hole.

DERIVATION OF SCHWARZSCHILD RADIUS

Now we'll derive the expression of Schwarzschild radius with some simple mathematics. It is very easy to understand. But, at first, we have to know about escape velocity (the term I already mentioned before).

Escape velocity is defined to be the minimum velocity an object must have in order to escape the gravitational field of the earth or any other celestial object. When we throw an object from the earth surface with its escape velocity, the object will never fall back. Simply, at escape velocity, the object's kinetic energy is equal to its potential energy.
If the mass of the object (which to be thrown) is m and velocity v, then kinetic energy will be $\frac{1}{2}mv^2$.

If the mass of the massive body is M, G is gravitational constant and R is the radius of the massive object.
Then for escape velocity (v_e),

$$\frac{1}{2}mv^2 = \frac{GMm}{R}$$

$$v_e = \sqrt{\frac{2GM}{R}}$$

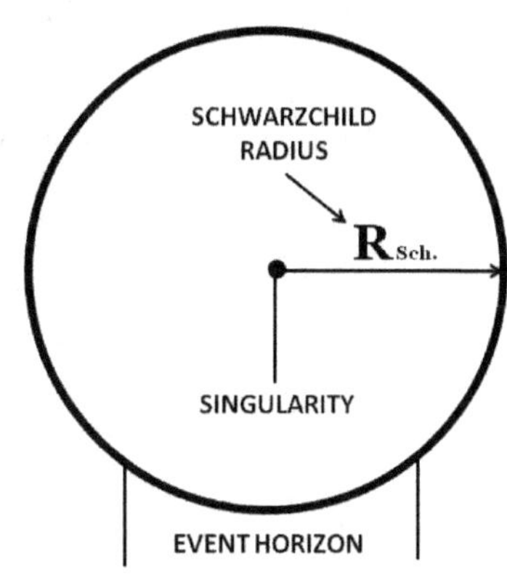

We know the escape velocity of a black hole is greater than or equal to the speed of light. So, $v_e = C$

Now,

$$C = \sqrt{\frac{2GM}{R_S}}$$

If R_S is Schwarzschild radius, then,

$$R_S = \sqrt{\frac{2GM}{C}}$$

It is the expression of the Schwarzschild radius.

The following table gives the Schwarzschild radii of some familiar astronomical objects:

OBJECT	MASS	SCHWARZCHILD RADIUS
SUN	2.0×10^{30} kg	3 km
EARTH	6.0×10^{24} kg	8.7 mm
MOON	7.3×10^{22} kg	0.11 mm
NEUTRON STAR	2.8×10^{30} kg	4.2 km

FORMATION OF BLACK HOLES

Do you remember the dialogue of Professor Albus Dumbledore from the 'Harry Potter and the chamber of secrets' movie? Where he says, "...Fox is a phoenix Harry, they burst into flame when it is time for them to die and then they are reborn from the ashes."
Similarly, a black hole starts its life with death.

Fusion is the process that powers the sun and the stars. It is the reaction in which two atoms of hydrogen combine together, or fuse, to form an atom of helium. In the process, some of the mass of the hydrogen is converted into energy. What happens to a star during the rest of its life depends on how massive it is. At its birth, a Star like our sun is in a delicate balance between its gravity which wants to make the star collapse onto itself and the pressure that pushes outwards that comes from the energy that's been produced during these fusion reactions happening in its core. However, at some point in the future, the hydrogen runs out, at this point of time, the core of the star will start to collapse in on itself under its own gravity. It gets denser and hotter until a point is reached where it can actually start to use the helium atoms themselves as the fuel for the fusion combining the helium atoms together and making carbon and oxygen. As the star begins to fuse the helium it creates more energy and this tremendous energy causes the outer layers of the star to expand. What happens is that the outer layers of the star get farther and farther from the centre. The force of gravity that the layers experience gets weaker and weaker. The atmosphere drifts off out into space and expands out to become a planetary nebula and it is one of the most beautiful objects in the universe.

Once the outer layers have drifted away, all that is left of the star is it's called a white dwarf. It is an incredibly dense object. It's dead! There's no nuclear fusion going on there anymore. It's incredibly hot. But then over millions of years, it will gradually cool down to become a black dwarf. Some stars, however, are much more massive than the sun. They're able to fuse into heavier and heavier elements inside their core. The star gets bigger and bigger. Some grow up to 1000 times the size of our sun until it has fused elements all the way up to iron. And once it formed an iron core there's no more energy left can be got from Fusion. Then it collapses. The rest of the star starts to collapse further, but then it triggers a huge shockwave within a quarter of a second. The outer parts the star of blasts off into space in a huge explosion known as "Supernova". For the supergiant stars, all that is left is the super-dense core known as a neutron star an object that can have a mass greater than our sun but be less than 20 kilometres across. Neutron stars are made of neutrons in a highly compressed and condensed state. If the gravity is too strong for the neutrons then the collapse just carries on until eventually that whole core has been compressed to a point and that is the black hole from which not even light can escape.

There are now thought to be four main types of black holes if classified by mass:

Primordial Black Holes have masses comparable to or less than that of the Earth. These purely hypothetical objects that might have formed through the gravitational collapse of regions of high density at the time of the Big Bang.

Stellar Mass Black Holes have masses between about 4 and 15 solar masses and result from the core-collapse of a massive star at the end of its life.

Intermediate Mass Black Holes of perhaps a few thousand solar masses may also exist. Sketchy evidence suggests that they may be found in some clusters of stars, and may eventually grow into supermassive black holes.

Supermassive Black Holes weigh between 106 and 109 solar masses and are found at the centres of most large galaxies.

Black holes can also be classified according to their two other properties of rotation and charge:

Schwarzschild Black Hole, otherwise known as a 'static black hole', does not rotate and has no electric charge. It is characterised solely by its mass. Kerr Black Hole is a rotating black hole with no electrical charge. Charged Black Hole can be of two types. A charged non-rotating black hole is known as a Reissner-Nordstrom black hole, a charged, rotating black hole is called a Kerr-Newman black hole. A binary black hole is a system consisting of two black holes in close orbit around each other.

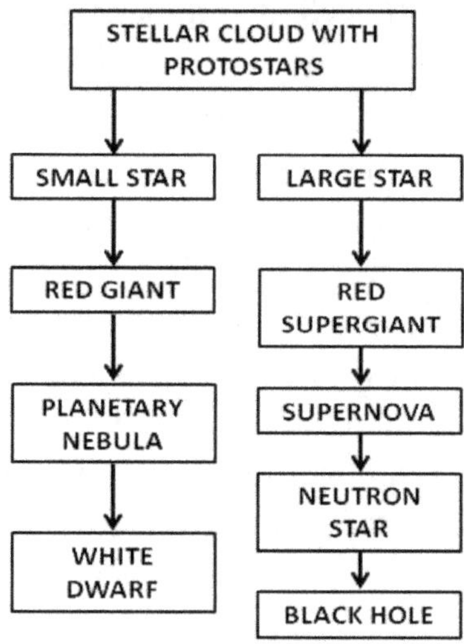

CHANDRASEKHAR LIMIT

Subrahmanyan Chandrasekhar was an Indian born American astrophysicist who spent his professional life in the United States. He was born on 9 October 1910 at Lahore, Punjab, British India. 

He was awarded the 1983 Nobel Prize for Physics with William A. Fowler for "theoretical studies of the physical processes of importance to the structure and evolution of the stars". His mathematical treatment of stellar evolution yielded many of the best current theoretical models of the later evolutionary stages of massive stars and black holes. The *'Chandrasekhar limit'* is named after him.

White dwarf stars are the end products of the stellar evolution of low to medium mass stars like our Sun. They are extremely dense objects (1 teaspoon of white dwarf material would weigh several tonnes!) and are supported against further gravitational collapse by electron degeneracy pressure. The Chandrasekhar Limit of 1.4 solar masses, is the theoretical maximum mass a white dwarf star can have and still remain a white dwarf (though this limit does vary slightly depending on the metallicity). Above this mass, electron degeneracy pressure is not enough to prevent gravity from collapsing the star further into a neutron star or black hole. His calculations made people understand about supernovas, neutron stars and black holes (which was identified in 1972).

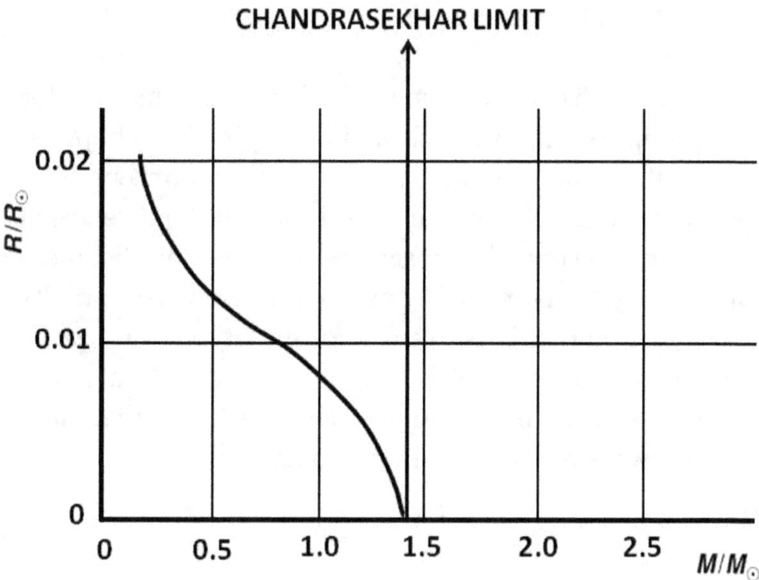

Figure 5.5: Chandrasekhar showed that the mass of a white dwarf could not exceed 1.44 times that of the Sun. Back then, it was thought that white dwarfs would be the culminating point for all stars.

GRAVITATIONAL TIME DILATION

We already discussed time dilation phenomena (for moving objects time runs slow) in the previous chapters. But, gravitational time dilation is a different concept. This effect measures the amount of time that has elapsed between two events by observers at different distances from a gravitational mass. Time runs slower where gravity is strongest, and this is because of gravity curves space-time. Does it seem a little complex to you? Okay, let's discuss this in a more simple way. We know that light speed is always constant and we also know that,

$$speed = \frac{distance}{time}$$

So, $\quad time = \frac{distance}{speed}$

Now, two beams of light, one in a *weak* gravitational field travelling between points A and B, and the other in a *strong* gravitational field travelling between points C and D. The path between c and d is longer due to relatively more curvature of space-time near strong gravitational field than near weak gravitational field. We know that light takes more time for a more curved path to reach D from C because here CD distance is greater than AB.

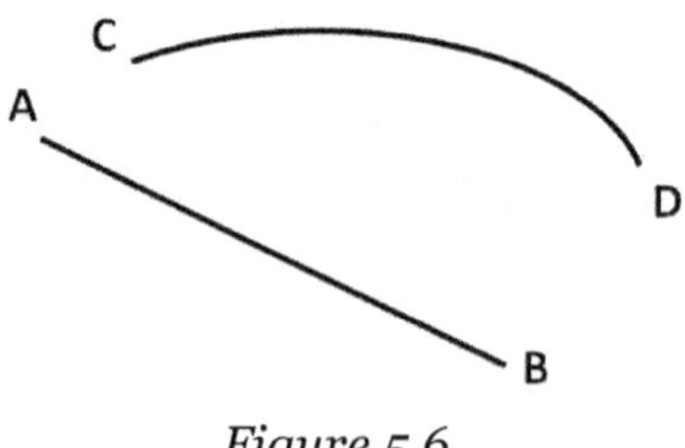

Figure 5.6

Now take another scenario, where you have two clocks, one is placed closer to the surface of the earth while the other far away from the gravitational pull of earth i.e. at some height above the earth's surface. The clock which is closer to the surface of the earth will run slower than the clock placed high up in space. This happened also due to gravitational time dilation. For example, considered over the total time-span of Earth (4.6 billion years), a clock set at the peak of Mount Everest would be about 39 hours ahead of a clock set at sea level. We all know about the Global Positioning System (GPS). The GPS satellites are positioned about 12,550 miles above Earth's surface and therefore are not as close to Earth's gravitational field. The clocks on these satellites tick faster than the clocks on Earth's surface so scientists have put a correction into the satellite programs to ensure that the GPS data sent back to Earth's surface have matching times. Without this correction, GPS satellites would not be so useful device since time synchronization is essential for the data sent by the GPS satellite.

The first proof for gravitational time dilation came when highly accurate atomic clocks were put in airplanes and after being flown around at high altitudes. Those clocks were synchronised before the plane journey. After the journey, the atomic clock on the plane was compared to another on the ground. Then we can notice an interesting thing, the clock on the plane was found to be faster than the one on the ground by the amount predicted by gravitational time dilation.

Gravitational time dilation has also been confirmed by the Pound–Rebka experiment (1959), observations of the spectra of the white dwarf Sirius B, and experiments with time signals sent to and from Viking 1 Mars Lander.

Interstellar is a science-fiction film, directed by Christopher Nolan. In the movie, Coop's team decides to land on Dr. Miller's ocean planet. It's orbiting Gargantua, a supermassive black hole that exists in the foreign galaxy, Every hour on that planet is seven years on Earth due to Gargantua's massive gravitational pull.

Relativity theory (GRT) proposed that the larger the object, more is the space-time is warped and twisted. So, Gargantua warps space-time; its gravitational strength bends the space-time around Miller's planet. This movie was a perfect presentation of gravitational time dilation near a supermassive black hole.

Kip Thorne, a theoretical physicist, served as the scientific consultant and executive producer of this movie.

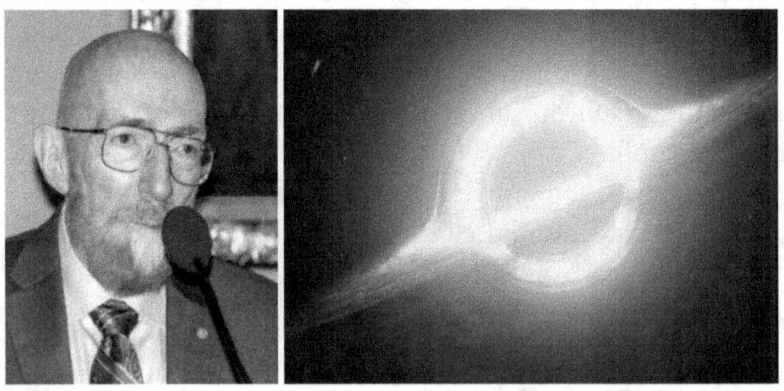

Figure 5.7: Kip Stephen Thorne is an American theoretical physicist and Nobel laureate, known for his contributions in gravitational physics and astrophysics (left).
The Gargantua black hole from the Interstellar movie (right).
(Picture credit: Double Negative)

JOURNEY INTO A BLACK HOLE

Let's take the example of two persons, you and your friend. Let's think that you fall towards a black hole. you will move into it faster and faster, accelerated by its gravity. Your feet will feel a stronger gravitational pull than your head because they are closer to the black hole. These are called 'tidal forces'. Your elongated body continues to stretch lengthwise and continue to shrink widthwise. Then your body is stretched apart. Your body would be stretched toward the singularity. This stretching is so incredible and so strong that your body is completely torn apart before you reach the event horizon. Scientists call it 'spaghettification' (comes from the word 'spaghetti', a long, thin, solid, cylindrical pasta.) If you fall into a supermassive black hole, your body remains intact, even as you cross the event horizon. But soon thereafter you reach the central singularity, where you are squashed into a single point of infinite density. You have become one with the black hole. Once you reach this point, you would sadly die! (see figure 5.8).

Singularity is a point where the laws of physics do not work. It is one of the most bizarre things in our universe. We cannot know what exactly happens at the singularity. Now consider, your friend who was standing at a safe distance from the gravitational effect of the black hole, watching you jump into the black hole. Your friend will experience your falling into the black hole completely differently! He would see your approach become slower and slower until you reached the event horizon. Now a very interesting thing would happen. He will always see you near the event horizon. It would seem to him that you are almost frozen in space. The light coming from your body becoming increasingly redshifted until you simply faded into nothingness and suddenly you will vanish for your friend (figure 5.9 illustrate this scenario).

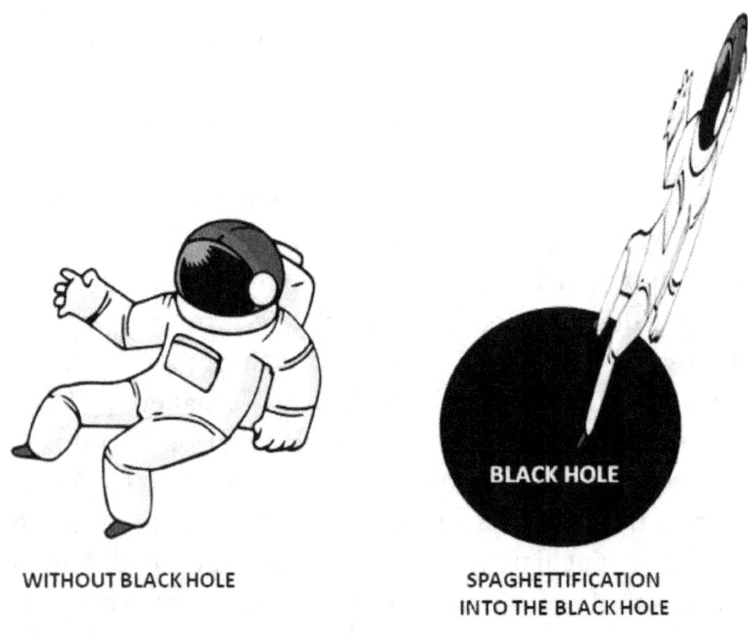

Figure 5.8

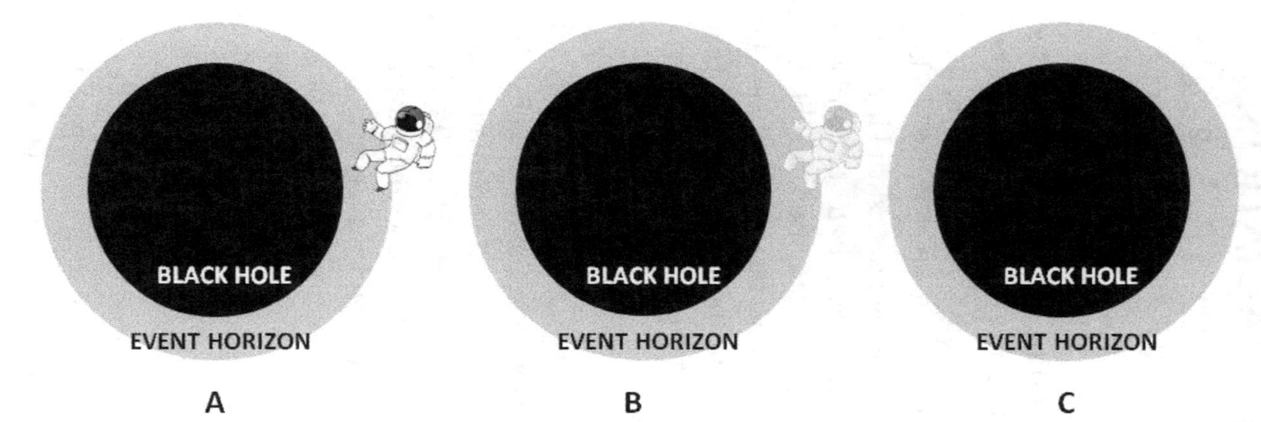

Figure 5.9

In figure A, You would seem almost frozen in space at event horizon. In figure B, your body becoming more indistinct due to redshift of light coming from your body. In figure C, you will suddenly vanish for your friend.

HAWKING RADIATION

Now a very interesting question is, *do Black Holes Die?* Or it will remain forever. Because once formed, there was nothing in theory or imagination that could bring material consumed back to the outside universe. These black holes should exist forever, only growing never shrinking. But no! Black holes have a finite lifetime due to the emission of Hawking radiation. Hawking radiation is black-body radiation that is predicted to be released by black holes, due to quantum effects near the event horizon (named after the famous theoretical physicist Stephen W. Hawking).

Before 1974, physicists believed that Black hole is totally dark and invisible. It is a state of ultimate darkness. But in 1974, a young theoretical physicist Stephen Hawking published a paper in Nature entitled 'Black Hole explosions' and in a follow up 1975 paper, he gave a theory combining quantum mechanics and the general theory of relativity. He started his calculations with rotating black holes. But when Hawking did the calculation, he found that even non-rotating black holes produce radiation. This type of radiation is known as Hawking radiation.

We know about the classical vacuum, which is devoid of everything. From Heisenberg's uncertainty principle we know that energy in empty space can never be zero. There is a state called quantum vacuum, which is also empty but this empty space seethes with activity as pairs of virtual particles matter and antimatter spontaneously appear and then annihilate each other briefly borrowing energy from the vacuum itself. This process is called quantum fluctuation.

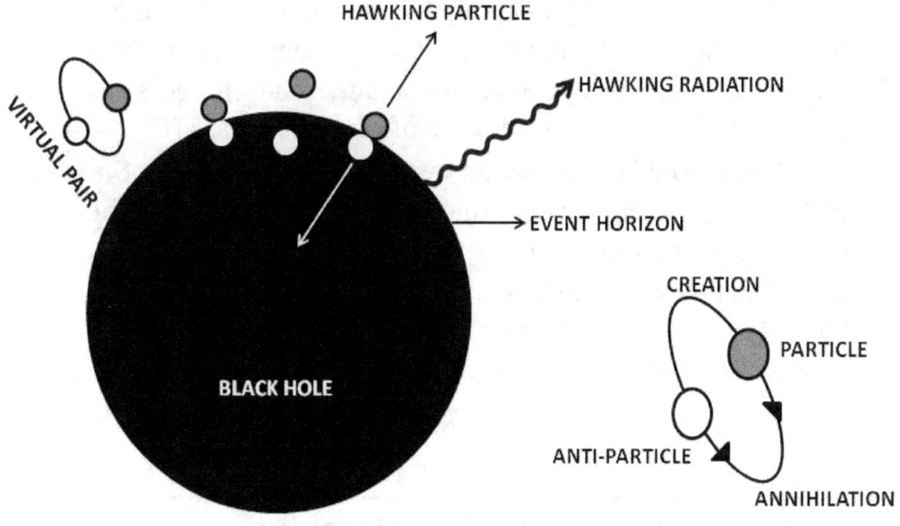

Figure 5.10: Process of Hawking radiation

But when this happens near the event horizon of a black hole, sometimes one of the pairs will be seized by the event horizon leaving the other virtual particle free to escape and taking its stolen energy with it. That energy cannot come from nothing and so the black hole itself pays the energy debt by slowly leaking away its mass. Because, according to Einstein's equation, $E = mc^2$ (where E is the energy, m is mass and c is the speed of light), which implies energy is proportional to mass. Now, as the mass decreases, the temperature increases due to the equation

$$T = \frac{hc^3}{16\pi^2 GkM}$$

Here h is the plank's constant. If the temperature is high, the black hole may radiate and lose its mass faster until it evaporates away entirely. However, the process is extremely slow. For a black hole of one solar mass ($M_\odot$ = 1.988 92 × 10^{30} kg), we get an evaporation time of 2.098 × 10^{67} years, much longer than the current age of the universe at (13.799 ± 0.021) × 10^9 years. So, theoretically, the black holes remain forever in our universe, because our life seems to be infinitely small compared to the evaporation time of black holes.

Stephen William Hawking

Professor Stephen William Hawking was born on 8th January 1942 (exactly 300 years after the death of Galileo) in Oxford, England. At the age of eleven, in 1952, Stephen went to St. Albans School and then on to University College, Oxford.

He wanted to study mathematics but mathematics was not available at University College, so he pursued physics instead.

In October 1962, Stephen arrived at the Department of Applied Mathematics and Theoretical Physics (DAMTP) at the University of Cambridge to do research in cosmology. In 1963 Stephen was diagnosed with ALS, a form of Motor Neurone Disease, shortly after his 21st birthday. He did PhD with his thesis titled 'Properties of Expanding Universes' in 1965. In 1966 he won the Adams Prize for his essay 'Singularities and the Geometry of Space-time'. Professor Stephen Hawking worked on the basic laws which govern the universe. In 1970, he showed that Albert Einstein's general theory of relativity implied space and time would have a beginning in the Big Bang and an end in black holes. In 1974, he discovered that black holes should not be completely black, but rather should emit 'Hawking radiation' and eventually evaporate and disappear. Stephen Hawking's most popular books are his best seller *A Brief History of Time*, *Black Holes and Baby Universes and Other Essays*, *The Universe in a Nutshell*, *The Grand Design* and *My Brief History*. Professor Stephen Hawking received thirteen honorary degrees. He was awarded CBE in 1982, Companion of Honour in 1989 and the Presidential Medal of Freedom in 2009. He was the recipient of many awards, medals and prizes, most notably the Fundamental Physics prize (2013), Copley Medal (2006) and the Wolf Foundation prize (1988). He was a Fellow of the Royal Society and a member of the US National Academy of Sciences and the Pontifical Academy of Sciences. Professor Stephen Hawking died at his home in Cambridge, England, on 14 March 2018, at the age of 76.

DETECTION OF BLACK HOLES

We know that black holes cannot be seen as it does not emit light. But we can detect its presence by measuring its effect on objects around it. There are three most significant effects. One of them is 'mass'. At first, we have to notice the behavior of the objects near it and then we have to notice their movements. Then use measurements of the movement of objects around a suspected black hole to calculate the black hole's mass. For example, in the core of galaxy NGC 4261, there is a brown, spiral-shaped disk that is rotating. The disk is about the size of our solar system but weighs 1.2 billion times as much as the sun. Such a huge mass for a disk might indicate that a black hole is present within the disk. Another important effect is the 'gravitational lensing effect' (which we already discussed in the previous chapter). The third most important effect is the radiation coming out from a black hole. Materials near a black hole can fall into a black hole and get heated to millions of degrees. These superheated materials emitted X-rays. We can detect it by X-ray telescopes like Chandra X-ray observatory. Besides X-rays, some black holes also emit jets. As for example, the star Cygnus X-1 is a strong X-ray source and it is the first such source widely accepted to be a black hole. It was discovered in 1964. It belongs to a high-mass X-ray binary system, located about 6,070 light-years from us. The primary star, HDE 226868, is a hot supergiant revolving about an unseen companion with a period of 5.6 days. The companion has a mass greater than seven solar masses. A star of that mass should have a detectable spectrum, but the companion does not emit any. So it is thought be a black hole.

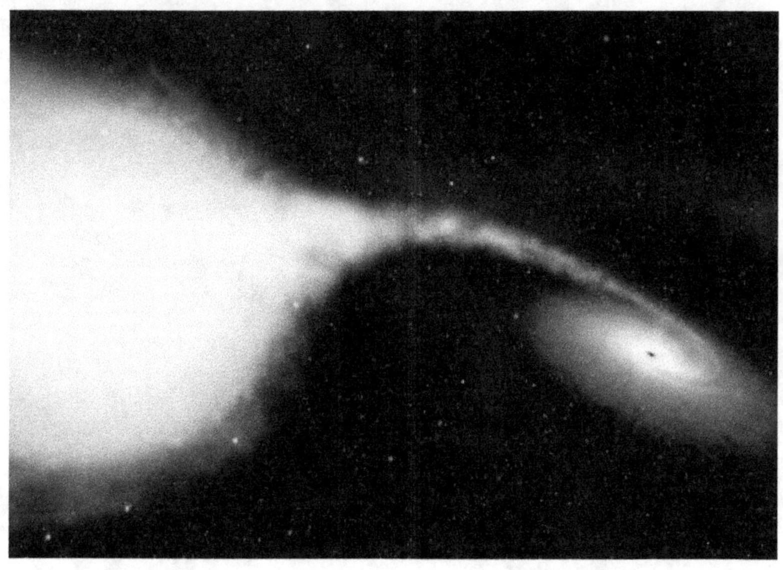

Figure 5.11: Artist's impression of Cygnus X-1

Picture courtesy: NASA, ESA, Martin Kornmesser (ESA/Hubble)

CHAPTER 6
PHOTOGRAPHY
OF THE INVISIBLE

"The history of astronomy is a history of receding horizons."

Edwin Hubble

There is a branch in Astronomy where we can study and detect celestial objects at radio frequencies, called 'Radio astronomy'. It uses large radio telescopes and antennas by which we can detect more objects than regular optical telescopes. These can identify different sources of radio emission. In the 1860s everyone knows about electromagnetic waves after the discovery of James Clark Maxwell. Many attempts were made to detect radio emission from the Sun at that time but these were unable to detect any emission. There were mainly two reasons: first is the technical limitation of instruments at that time and another was the effect of the Ionosphere. In 1902, after the discovery of the Ionosphere, physicists know that this layer can reflect any astronomical radio waves back into space. In 1932 Karl Jansky first observed the radio waves coming from the Milky Way. At first, there was only dish type radio telescope. An American scientist Grote Reber first built a parabolic radio telescope in 1937 and conducted the first sky survey by radio frequencies.

Figure 6.5: In a side yard of his mother's house in Wheaton, Illinois, a 26-year old engineer named Grote Reber built the first dish antenna radio telescope in 1937. He used wooden rafters, galvanized sheet metal, and spare parts from a Ford Model T truck. With this 31-foot diameter telescope, Reber mapped the radio structure of our Galaxy, discovered bright sources of radio waves outside our Galaxy, and made the observations that would later help physicists discover non-thermal radiation.

(Picture courtesy: NRAO/AUI/NSF)

In 1942, a British Army research officer, James Stanley Hey first detected the radio waves emitted by the Sun.
Radio astronomy has various techniques to observe objects in the radio spectrum. Radio telescopes are antennas and

radio receivers used to receive radio waves from astronomical radio sources in space. In the first chapter, we have seen various frequency zones of the electromagnetic spectrum (see figure 1.2). In optical telescopes, we use the optical or lightwave portion of the electromagnetic spectrum but here, in radio telescopes, we use the radio frequency part of the electromagnetic spectrum. We can use one single radio telescope or many radio telescopes connected through cables, optical fibres. In the second case, we can get pictures with high resolutions. To achieve more high resolution with a single radio telescope, we need 'radio interferometry'. This technique was developed by British radio astronomer Martin Ryle and Australian engineer, radio astronomer Joseph Lade Pawsey and Ruby Payne-Scott in 1946. Modern radio interferometers consist of many radio telescopes connected together. It increases total signals collected and increase high resolution.

Before April 10, 2019, we only had an imaginative picture of a black hole. This day was a very important day in the history of physics. On that day the first-ever image of a black hole was 'clicked'. The best part of this is that we got a picture of what we expected from Sir Einstein's theory of relativity. It was not an easy process because we can't observe a black hole through normal telescopes as light can't escape from it. But we know that black hole radiates and we can synchronize the recorded data from the readings of that radiation. But for that, we need 'Very-long-baseline interferometry' (VLBI).

VERY LONG BASELINE INTERFEROMETRY (VLBI)

VLBI means Very Long Baseline Interferometry and associated with radio astronomy and geodesy*. It is a type of astronomical interferometry used in radio astronomy. Here observations of an object that are made simultaneously by many radio telescopes to be combined, emulating a telescope with a size equal to the maximum separation between the telescopes (as for example, European VLBI Network spans over three continents with max. baseline of 10000 km). VLBI has mainly three categories: First is Continental, where the baselines of 100s to 1000s of km; second is Global, where baselines are 1000s km or more and third is space VLBI involving the use of satellites. So simply, we can use the whole earth as one telescope by this method. But the earth's size limits the baseline lengths below 12000km. We can also increase the resolutions by sending an antenna to space. Data received at each antenna in the array include arrival times from a local atomic clock. The data are correlated with data from other antennas that recorded the same radio signal, to produce the resulting image.

VLBI has its application for imaging distant cosmic radio sources, spacecraft tracking, and for applications in astrometry.

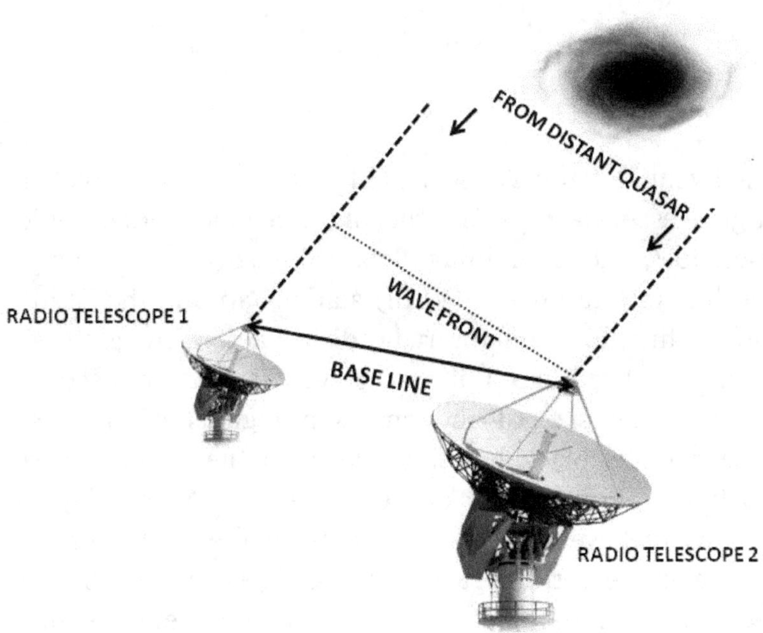

Figure 6.2: This is a simple representation of the VLBI process

---

* Geodesy is the Earth science of accurately measuring and understanding Earth's geometric shape, orientation in space, and gravitational field. The field also incorporates studies of how these properties change over time and equivalent measurements for other planets. It is a branch of applied mathematics.

EVENT HORIZON TELESCOPE

The Event Horizon Telescope (EHT) is a project to create a large telescope array consisting of a global network of radio telescopes and combining data from several very-long-baseline interferometry (VLBI) stations around the Earth. This technique of linking radio dishes across the globe to create an Earth-sized interferometer has been used to measure the size of the emission regions of the two supermassive black holes. It observed the supermassive black hole Sagittarius A* at the centre of the Milky Way, as well as the even larger black hole in the centre of the supergiant elliptical galaxy Messier 87, with angular resolution comparable to the black hole's event horizon. The EHT collaboration involves more than 200 researchers from Africa, Asia, Europe, North and South America. The international collaboration is working to capture the most detailed black hole images ever by creating a virtual Earth-sized telescope (Figure 6.2).

Data collected on hard drives from the various telescopes and sent to MIT Haystack Observatory in Massachusetts, USA, and the Max Planck Institute for Radio Astronomy, Bonn, Germany, where the data are cross-correlated and analyzed on a grid computer made from about 800 CPUs all connected through a 40 Gbit/s network. Once the EHT has measured data from the black hole, we have to make a picture from it by a process called 'imaging'. It gives us some indication of the structure of the black hole. We are still missing some information about the black hole's image. The imaging algorithms we develop fill in the gaps of data we are missing in order to reconstruct a picture of the black hole (figure 6.4).

On April 10, 2019, the first image of the black hole inside galaxy Messier 87 was published. The image shows a bright ring formed as light bends in the intense gravity around a black hole that is 6.5 billion times more massive than the Sun (figure 6.5).

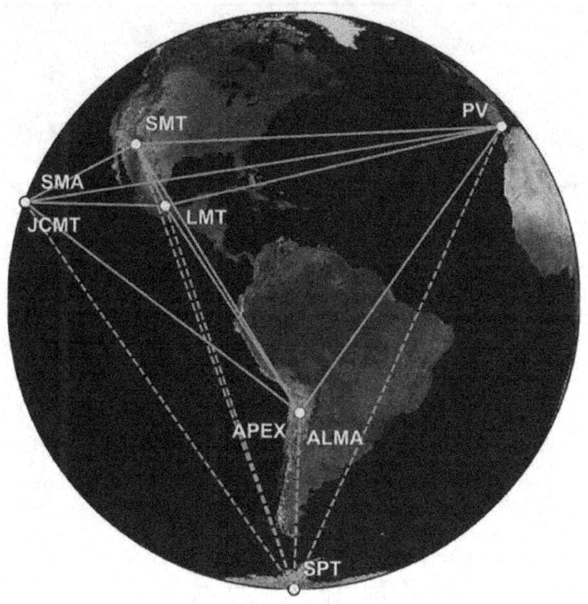

Figure 6.3: EHT Collaboration
(Picture courtesy: EHT collaboration / Astrophysical journal letters,
875(2019) L1/CC BY 3.0)

RADIO TELESCOPE 1
RADIO TELESCOPE 2
RADIO TELESCOPE 3
DIGITAL CONVERSION
DIGITAL CONVERSION
DIGITAL CONVERSION
ATOMIC CLOCK
ATOMIC CLOCK
ATOMIC CLOCK
HARD DRIVES
HARD DRIVES
HARD DRIVES
SYNCHRONIZATION PROCESS
DATA REDUCTION
IMAGE

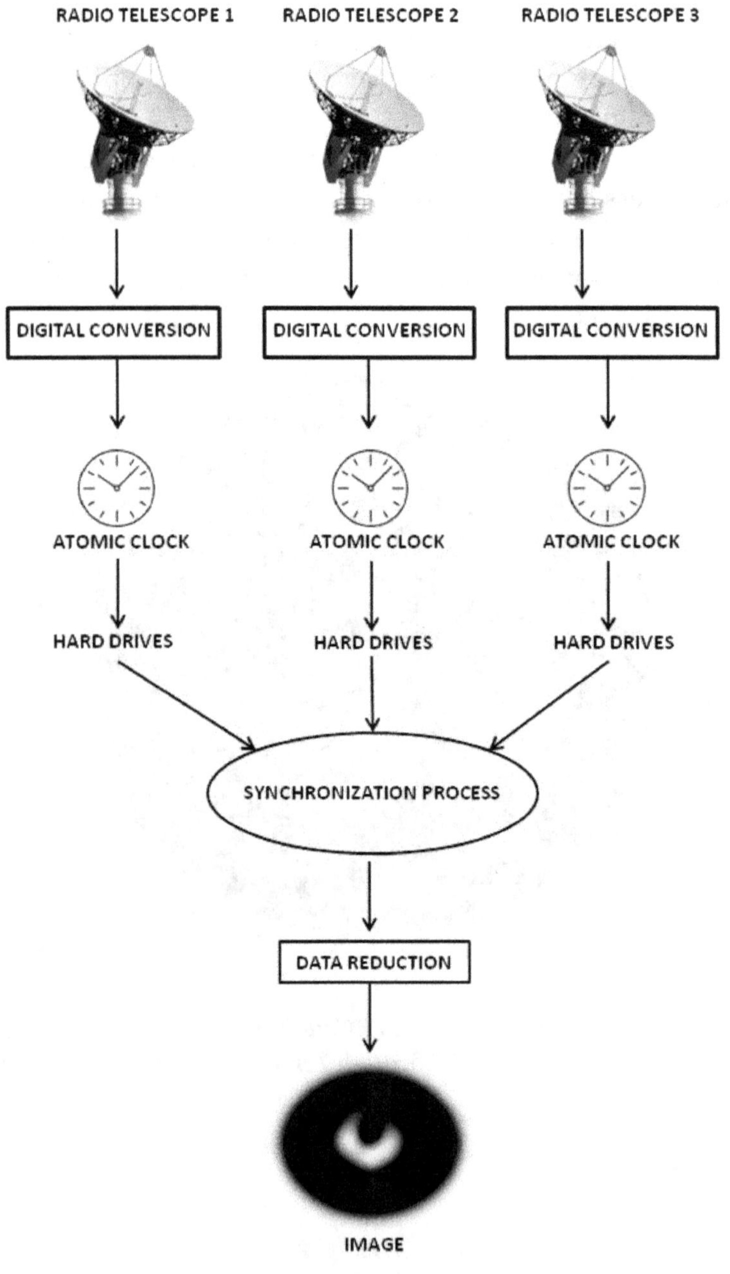

Figure 6.4

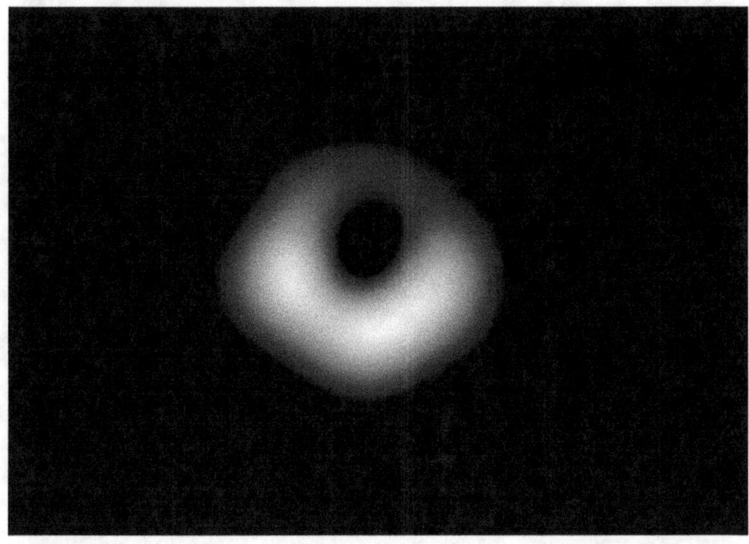

Figure 6.5: The first image of the black hole inside the galaxy Messier 87
(Picture courtesy: EHT Collaboration)

CHAPTER 7
THE REVERSE BLACK HOLE
WHITE HOLE

"Even if it turns out that time travel is impossible, it is important that we understand why it is impossible."

Stephen Hawking

One of the most interesting laws of nature is equivalence. If there are some good qualities in us then there should always be some bad qualities also. If there is light, there must be dark. There always be a bad part with a good part in us. We can always assume an opposite or reverse condition against every phenomenon. According to this, we can also think about the opposite condition of black holes; and yes, there it is – the White Hole! We can predict this from the solutions of Einstein's equations of the general theory of relativity.

Let, assume a situation like the reverse of a black hole. In which nothing can be entered from the outside but from which matter and light can escape. This is called 'White Hole'. White holes are the completely hypothetical mathematical concept. They attract matter but objects falling towards it would never reach its event horizon. Mathematically it is the simplest kind of black hole, with no existence mass. So, a white hole is something that probably cannot exist in the real universe. White holes are a possible solution to the laws of Albert Einstein's general theory of relativity (mainly Einstein's field equations). The possibility of the existence of such an object was first proposed by a Russian cosmologist Igor Novikov in 1964. But the question is why white holes cannot exist?

Here the main problem is that white holes decrease entropy. This is the direct violation of the second law of thermodynamics. In this universe, we obey the laws of thermodynamics. According to thermodynamics, the net entropy of the universe is always increasing, it cannot be decreased. In the case of white holes, we know that nothing can enter it but objects always escape from it. It means white holes create these objects. For this, the entropy of this isolated system decreased. So, according to physics white hole cannot exist or if it exists, it must be for a few seconds for the instability. But, as it accomplished the Einstein's field equation, we should study it. Let's see what happened if it actually exists. We know matter and energy cannot be created or destroyed. We also know matters can enter a black hole but cannot come out. So if we thought black hole exists forever then there is no problem. All matter kept securely in it. But according to Stephen Hawking, black holes emit radiation and it evaporates continuously. So, one day they should vanish. Now the problem arises. If black holes vanish then what happened to the matters inside it? as they cannot be destroyed. It is here that the concept of the white hole plays an important role. If we thought that at the end stage black hole converted into a white hole and emits all matter that the black hole ate, then the problem seems to be solved (look at figure 7.1). This concept also explains one most important question, how Big bang happened? Some researchers have proposed that the big bang may occur at the core of a white hole when it formed. This could create a new universe. The new universe exists outside of the parent universe.

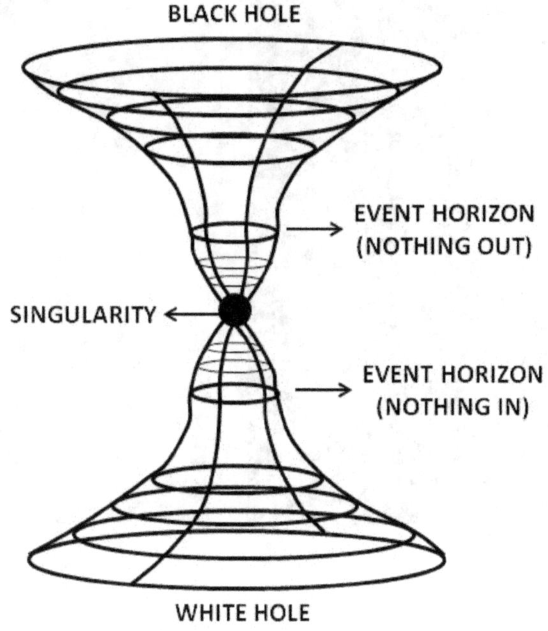

Figure 7.1

If this is true, then maybe our universe is a part of another parent universe which is much larger than ours.

Basically, black hole and white holes attached to each other as depicted in the above picture and it creates a portal through which one may travel one universe to other universes or different parts of the same universe. Yes, Sounds like something out of a science fiction movie. But it is known as '*Einstein-Rosen Bridge*' or more popularly as '*Wormhole*'.

Figure 7.2: Albert Einstein (left) and Nathan Rosen (right)

THE IDEA OF WORMHOLE

In 1935, Albert Einstein and Nathan Rosen used the general theory of relativity to elaborate the idea of worm hole, proposing the existence of bridges through space-time. Now, the question is, why is a wormhole called a 'wormhole'? or what is the significance of the term 'wormhole'? It is a very simple idea. This bridge is a shortcut through the fabric of space-time, just as a worm burrows through fruits.

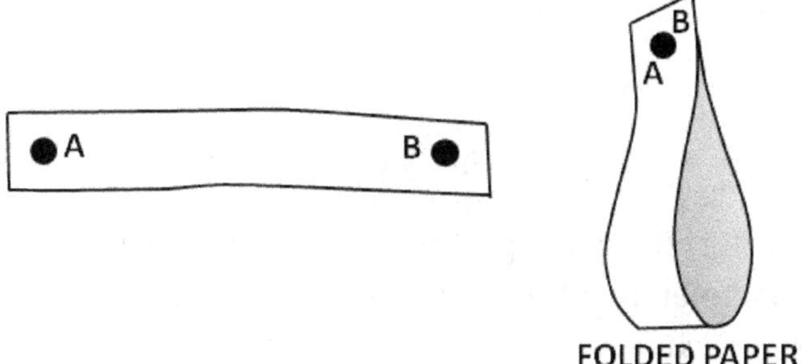

Figure 7.3

We can understand this concept by a piece of paper (see figure 7.3). Let, there are two points A and B on a piece of paper. Now tell me, what is the shortest distance between those points? What do you think? You may say the straight line connecting those two points should be the shortest distance. Yes, it is true. But, if we overlap those points by folding the paper then it will be the shorter distance than the previous answer. Technically the two points will be at the same point in space-time. Suppose the point A is your home and point B is your school (figure 7.4). Now, if you can fold the space-time such that point A and point B overlap to each other then the moment you step out of your home is the moment you step into your school. So, theoretically, the time taken for this is zero or negligibly small. It reminds me of the 'anywhere door' of Doremon cartoon (one of my favourite cartoon show).

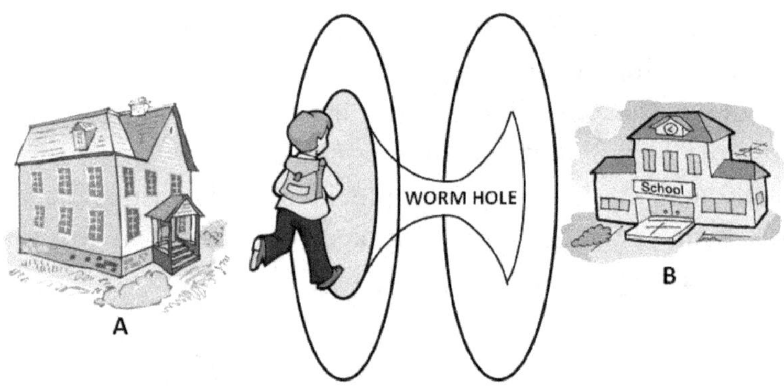

Figure 7.4

We can travel from one part of our universe to another part or to a completely different universe through worm hole in just a few seconds or less than that time. Einstein's theory of general relativity mathematically predicts the existence of wormholes, but none have been discovered to date.

We can see so many scenes of time travel through worm hole in Science fiction movies and novels. But the reality of that process is more complicated. The concept of worm hole still a hypothetical concept just because we cannot spot them due to mainly two reasons.

The first problem is its size. The worm holes are predicted to exist on the microscopic level. So, it is very hard to detect and almost impossible to travel within it for us. But, as the Universe is expanding, some of them may stretch to the larger size. The second problem is more important. It is about their stability. These worm holes would be useless for space-time travel because they collapse very quickly. This problem can be solved by using 'exotic matter'*. Worm holes containing exotic matter are more stabilized. They could stay open for a longer period of time. If a worm hole has enough exotic matter or we artificially add sufficient exotic matter to it then it can be used for travel through space. Some researchers say that it also can be used for time travel. But, it will be challenging for a human. British cosmologist Stephen Hawking has argued that such use is not possible. There is one more problem. Even if we can stabilize the worm hole by adding exotic matter in it and travel through it, there is still a possibility that the addition of regular matter (here human beings) could destabilize the portal. We need the most advanced technology to detect and use worm holes as a time machine.

---

* Exotic matter is not dark matter or antimatter. It has 'exotic' properties. It is non-baryonic. It has negative energy density and negative pressure. Most of the exotic matter remains hypothetical. Tachyons are examples of exotic matter. The Tachyons are very interesting because according to the theory of relativity these particles are able to travel faster than the speed of light. Exotic matter can violate known laws of physics.

CHAPTER 8
DARK MATTER

DARK MATTER

"You know, dark matter matters"
Neil deGrasse Tyson

Now, in this chapter we will explore a deeper universe. The part of the Universe visible to us in current telescopes (with most advanced technology till now) only represents a very small fraction of the total amount present in the Universe. All the stars, planets, galaxies and other visible objects make up just 4.4 percent of the Universe (normal matter inside stars are about 0.6 percent and that outside stars are about 3.8 percent). The other 96 percent is made of completely mysterious, invisible substance. Astronomers have not yet observed those directly. It's completely invisible to light and other forms of electromagnetic radiation. But, they are very confident it exists because of the same effects. These invisible things are called 'Dark matter' (25 percent) and

'Dark energy' (70 percent). Remember, here 'Dark' means 'we don't know' or 'unseen'. Although 'Dark matter' and 'Dark energy' have similar-sounding names, they refer to two very different things.

Any non-luminous objects in the Universe can be detected by its gravitational influence on the luminous objects. For example, gravitational lensing, the motion of galaxies in galaxy clusters, hot gas in galaxy clusters etc. The first detection of dark matter was made in 1933 by Zwicky. According to physics, stars at the edges of a spinning spiral galaxy should travel slower than those near the galactic centre, where visible matter concentrated. But, observations show that stars orbiting at the almost same speed regardless of where they are in the galactic disk. Zwicky measured the velocity dispersion of galaxies in the Coma cluster and observe that effect. His work was followed up by Smith in 1936 for the Virgo cluster. This result can be solved if we assume that the boundary stars are feeling the gravitational effects of an invisible mass or there is something wrong with the theory of gravity. But, the known theory of gravity of Einstein solved so many problems of the Universe and it is so well tasted quite many times. So, if the theory of gravity is correct then there must be more matter than we can see.

Dark matter also can explain some optical illusions of deep Universe. We have known about gravitational lensing in the previous chapters. Pictures of galaxies include strange rings and arcs of light can be explained if we assume that the light from more distant galaxies is being distorted by invisible, massive clouds of dark matter. Even now, there's so much that we don't know about dark matter and dark energy. But, there are many things that we can say about them.

Figure 8.1: Mass distribution of galaxies
The total mass in dark matter is about 10 times more than in
stars

Let us try to understand that by which the dark matter made of. We can assume that dark matter made of WIMPs or MACHOs. MACHOs mean Massive Compact Halo Objects like free-rotating planets and brown dwarfs old white dwarfs that have ceased glowing, Neutron stars, Black holes etc. But, the problem is, some of the unseen mass was due to MACHOs, but not the majority of it. So, we can conclude that it should be WIMPs or Weakly Interacting Massive Particles. Dark matter is not dark clouds made by normal matter, matter made up by baryon particles. We can tell this because; we can detect baryon clouds by their absorption of radiation passing through them. Secondly, it is not antimatter. Antimatter is the normal matter with a few flipped properties. The sub-atomic particles of antimatter have properties opposite those of normal matter. As for example, antimatter of an 'electron' is a 'positron'. They both have the same mass but have the opposite electric charges. Antimatter was created along with matter after the Big Bang, but antimatter is rare in today's universe for some unknown reasons. There is a small amount of antimatter all over the space, even inside your body. Antimatter is a whole bunch of particles, called 'antiparticles'. The existence of antiparticle was predicted by Dirac in his relativistic quantum mechanical theory of electrons. Later, as predicted, antielectrons or positrons were discovered in cosmic rays. All elementary particles have a known antiparticle. Since every particle has its antiparticle, atoms and molecules can also be built up out of antiparticles. For example, ordinary H-atom has a proton in the nucleus and an extranuclear electron. If the proton is replaced by an antiproton and the electron by a positron, we shall get what may be called an atom of antihydrogen (figure 8.2). When antimatter comes in contact with its regular version of the matter, they mutually destroy each other and all of their mass is converted to energy. This process is called annihilation. Dark matter is not antimatter, because we do not see gamma rays that are produced when antimatter annihilates with matter.

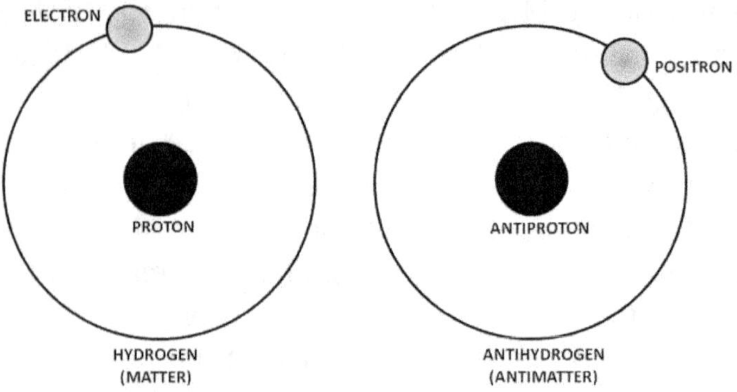

Figure 8.2: Idea of matter and antimatter

Dark matter is also not 'Neutrinos'. Neutrinos are almost massless, invisible, and intangible. They always are travelling close to the speed of light. We know that Dark matter determined the development of large-scale structure like galaxies and clusters of galaxies in the universe. If most of the dark matter were neutrinos, they wouldn't stay put long enough to let those structures form. So, we can conclude that dark matter made up of a completely new particle, unknown to particle physics. The nature of dark matter is one of the peculiar mysteries in physics.

Figure 8.3: This collage shows NASA/ESA Hubble Space Telescope images of six different galaxy clusters. The clusters were observed in a study of how dark matter in clusters of galaxies behaves when the clusters collide. 72 large cluster collisions were studied in total. Using visible-light images from Hubble, the team was able to map the post-collision distribution of stars and also of the dark matter (coloured in grey).
The clusters shown here are, from left to right and top to bottom: MACS J0416.1–2403, MACS J0152.5-2852, MACS J0717.5+3745, Abell 370, Abell 2744 and ZwCl 1358+62.
(Credit: ESA/Hubble)

CHAPTER 9
Doppler Effect

Whenever you are standing on the platform of a railway station, you have heard the increased pitch of an approaching train siren and the decrease in pitch as the train passes by.

This effect was first described by an Austrian physicist Christian Doppler in 1842 and is called the Doppler Effect. Doppler first proposed this effect in his treatise "*Über das farbige Licht der Doppelsterne und einiger anderer Gestirne des Himmels*" (*On the coloured light of the binary stars and some other stars of the heavens*). The Doppler Effect is the change in frequency or wavelength of a wave in relation to an observer who is moving relative to the wave source. This effect is observed whenever the source of waves is moving with respect to an observer. Doppler Effect can be observed for lightwave (electromagnetic waves) also. But, we are most familiar with the Doppler Effect of sound waves in our daily life. We don't notice the Doppler effect of light in daily life because light travels so much faster than the speed of the sound waves. At first, I discuss this effect for sound wave and then I will discuss the Doppler Effect for lightwave (which is a more important concept to understand some of the previous chapters in this book). To understand this effect for sound waves, at first, assume that both the source (in this case the train) and the observer (you) are at rest in the railway station as shown in figure 9.1. In this case, you will recognize the siren at the same frequency at which it is emitted. So, in this case there is no Doppler Effect. Now, in the second case, the train is moving towards you. Here you can hear a higher frequency sound of the siren. This case is illustrated in figure 9.2.

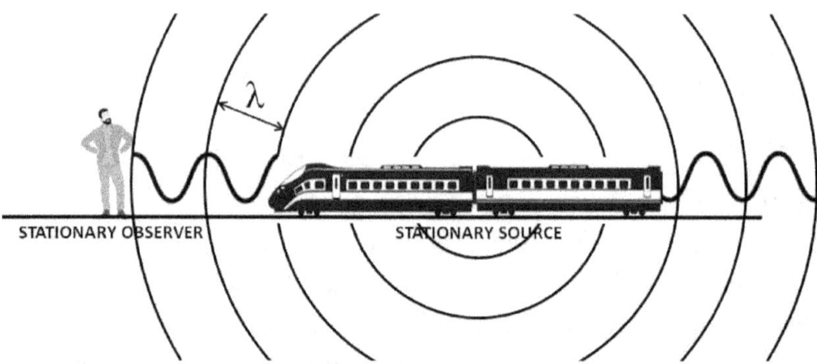

Figure 9.1: The circles represent the wavefronts from the siren's sound waves. In this case, you hear the same tone emitted by the train siren.

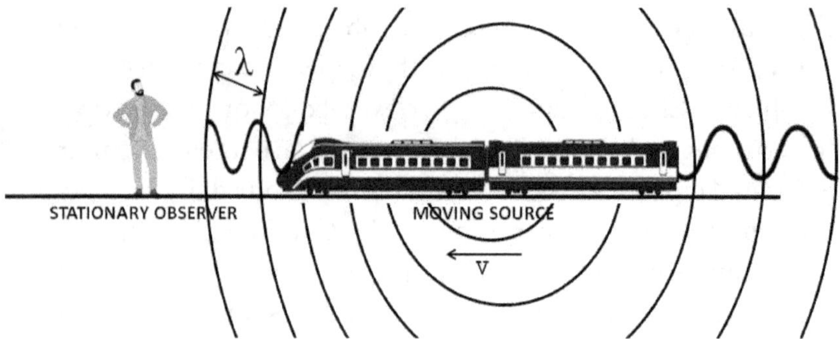

Figure 9.2: Here the wavefront contracted and the wavelength (λ) becomes smaller as the train moving towards you. The frequency becomes higher in this case as the wavelength and frequency are inversely proportional to each other. You will hear a higher pitch siren in this case.

In the third case, the train is moving away from you. Now, you can hear a lower frequency sound of the train siren. This case is illustrated in figure 3.

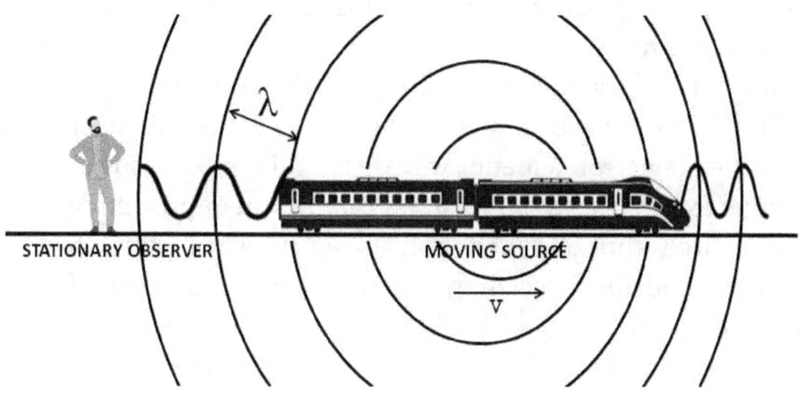

Figure 9.3: Here the wavefront stretched and the wavelength (λ) becomes higher as the train moving away from you. The frequency becomes lower in this case and you will hear a lower pitch siren in this case.

From these observations, we can formulate an equation for the relationship of source and observed frequencies in the Doppler Effect.

$$f' = \left(\frac{v \pm v_0}{v \mp v_s}\right) f_0$$

In this equation,
f' = Frequency of sound waves heard by the observer.
f_0 = Frequency of sound waves produced by the source.
v = Velocity of Sound in Air.
v_0 = Velocity of the observer.
v_s = Velocity of the source.

So, from here we can observe two more cases where you are moving away from the source and the source is also moving away from you, the frequency heard by you must decrease and oppositely if you are approaching a source which is moving towards you; in this case, the frequency heard by you increases.

I think you understand these examples. Now we take a step forward. Now we see what happens when we use light or any other electromagnetic waves in this effect. From previous examples, we understand that for an approaching wave source the observer notices an upward shift in frequency and for a receding source there is a downward shift in frequency noticed by the observer. Now we see the visible light region (VIBGYOR) of the electromagnetic spectrum. If we observe an upward shift of frequency, it is called Redshift because the light source appears more radish in colour. Oppositely, if we observe a downward shift of frequency then it will be called Blueshift. For redshift the light source approaching to us and for blue shift the light source receding from us. You can easily understand this phenomenon by figures 9.4.

We saw the figure of the electromagnetic spectrum in chapter one (Nature of light: Figure 1.2). The concept of redshift and blueshift can apply to any part of the electromagnetic spectrum. If radio waves are shifted into the ultra-violate part of the spectrum, they are said to be blue shifted (shift to higher frequency). Similarly, if the gamma-ray part is shifted into the ultra-violate part of the spectrum, they are said to be redshifted (shift to a lower frequency). The concept of redshift and blueshift is very useful in astronomy. But I will discuss the application of this concept after some time. Now, I discuss the *Relativistic Doppler effect* for light.

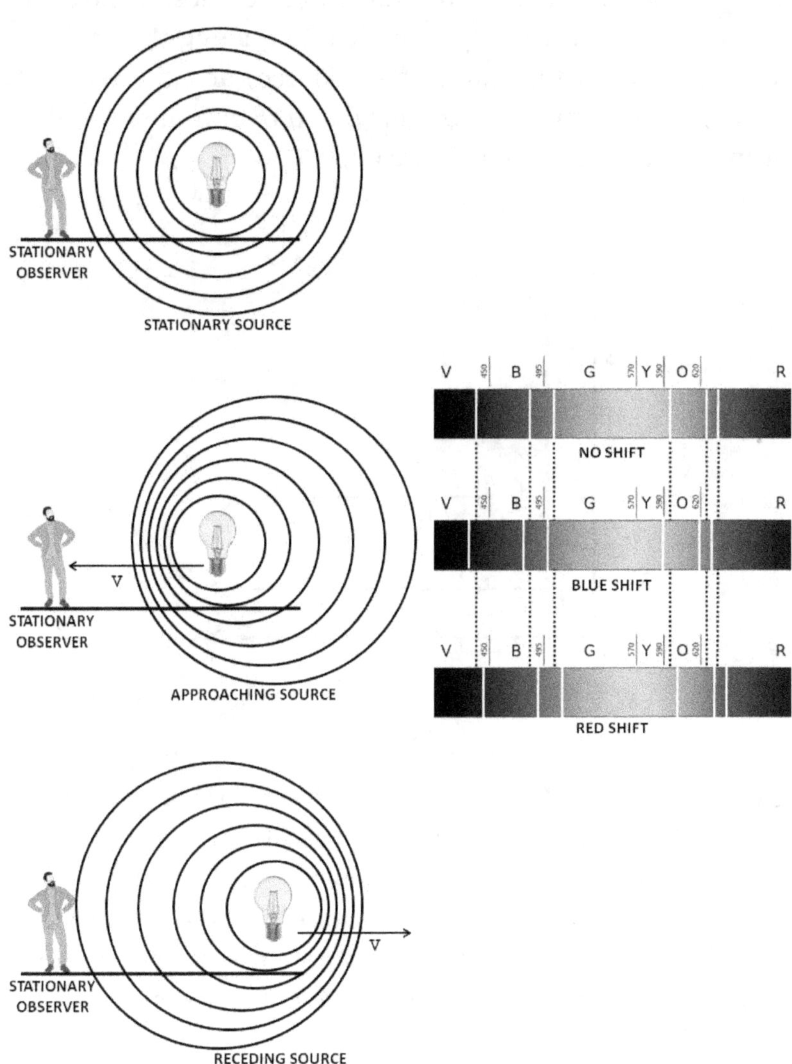

Figure 9.4: Here we see three cases for light source movement and the corresponding shift of lines (white) in the spectrum.

For the relativistic Doppler effect, the velocity of the light source should be near to the speed of light C. To understand it, consider two frames of references S and S'. S' is moving with a velocity v relative to S along the positive x-axis. Now, let the transmitter (source) and the receiver (observer) be situated at origins O and O' of frames S and S' respectively as shown in figure 9.5.

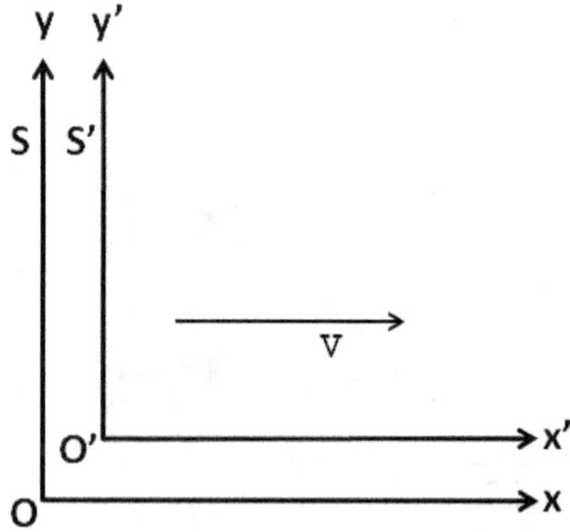

Figure 9.5

In this chapter, I directly write the relativistic Doppler effect's formulas without mathematical derivation. If you need mathematical derivations then see appendix 2 (page 151-154). If ϑ and ϑ' be the actual and observed frequencies of light pulses transmitted by the source or transmitter. Then after calculations, we can get a relation between them. This is,

$$\vartheta' = \vartheta \sqrt{\frac{\left(1 - \frac{v}{c}\right)}{\left(1 + \frac{v}{c}\right)}}$$

In terms of wavelength we get,

$$\lambda' = \lambda \sqrt{\frac{\left(1 + \frac{v}{c}\right)}{\left(1 - \frac{v}{c}\right)}}$$

Where λ is the actual wavelength and λ' is the observed wavelength. In these relations v is positive or negative according to the observer is receding from or approaching the source of light. Actually, this situation called *Longitudinal Doppler Effect* as the observations are made along the direction of travel of the light source. Another situation is called the *Transverse Doppler Effect* where the observations are made at the right angles to the direction of travel of the light source. The expression of Transverse Doppler Effect is,

$$\vartheta' = \vartheta \sqrt{\left(1 - \frac{v^2}{c^2}\right)}$$

If $\vartheta \ll c$ then we see that $\vartheta' = \vartheta$. So, according to classical theory, there is no transverse Doppler effect. The longitudinal Doppler effect was confirmed by H.E.Ives and G.R.Stilwell in 1941 using beams of hydrogen atoms in excited electronic states.

This redshift and blueshift are of intense interest to astronomers. They use this information in the frequency of electromagnetic waves produced by moving astronomical objects like stars, galaxies etc. The concept of redshift is very important for the next chapter which leads us to the concept of Dark energy.

CHAPTER 10
HUBBLE'S LAW

"Observations always involve theory."
Edwin Powell Hubble

In the previous chapter we know about the Doppler effect of light or electromagnetic spectrum and the concept of redshift (and blueshift). But what is the application of redshift and blueshift in science?

American astronomer Edwin Hubble was the first to describe the redshift phenomenon and used it to understand the concept of the expanding Universe.

After the publication of Albert Einstein's general relativity in 1915, Alexander Friedmann published a set of equations that were derived from general relativity, now known as Friedmann equations. Those equations first showed that the Universe might expand and this expansion has a calculable rate. Then Georges Lemaitre, in a 1927 article, independently derived that the Universe might be expanding. He observed the proportionality between recessional velocity (Recessional velocity is the rate at which an astronomical object is moving away, usually from the earth.) and distance to distant bodies, and suggested an estimated value of the proportionality constant. Later this constant was corrected by Hubble in 1929 and known as Hubble constant.

The Hubble-Lemaitre law, also known as Hubble's law states that the distances between celestial objects are continuously increasing and that therefore the universe is expanding. This is a statement of the direct relationship between the distance to an object and its recessional velocity as determined by the redshift. Mathematically this law can be written as

$$v = H_0 D$$

Here, v is the observed velocity of the astronomical object (say, a galaxy), usually in Km/Sec H is the 'Hubble's constant' in Km/Sec/Mpc. D is the distance to the galaxy in Mpc (1 Megaparsec equals to about 30.9 trillion or 3.09×10^{13} kilometres). Also the SI unit of H_0 is Sec-1 but it is mostly used in Km/Sec/Mpc. So, the reciprocal of Hubble's constant is the Hubble time.

In 1929, Hubble estimated the value of this expansion factor (H_0) to be about 500 Km/Sec/Mpc. But after many observations, scientists calculated the value as nearly 71 Km/Sec/Mpc ($\pm 10\%$).

However, obtaining the value for Hubble's constant is very complicated. While in general celestial objects follow the smooth expansion, the more distant ones moving faster away from us. This type of motions caused slight deviations from the line predicted by Hubble's law (figure 10.1). We can understand this motion with an easy explanation. If a galaxy A at a distant 1 Mpc away from the earth then it should receding from us at a rate 71 Km/Sec. But another galaxy B situated at a distance 2 Mpc from the earth then it should be moving away from us at a speed 142 Km/Sec which is twice the rate of galaxy A (figure 10.2).

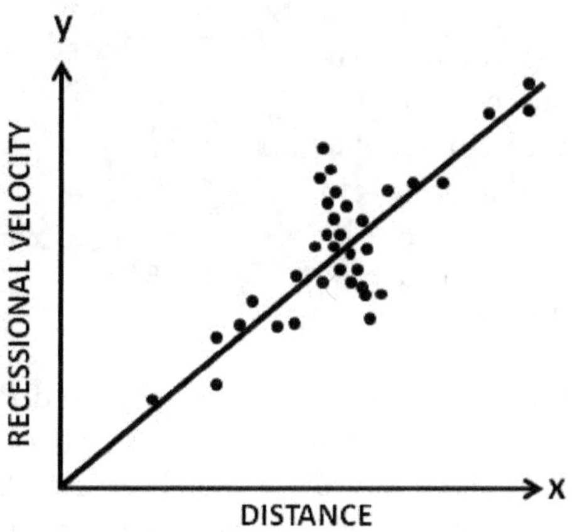

Figure 10.1: The recessional velocity of a few galaxies plotted against their distance from earth. On this graph, the slope of the line equals to Hubble's constant H_0

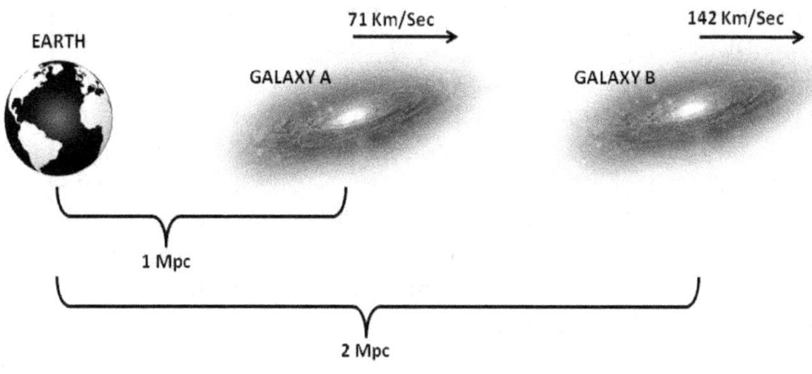

Figure 10.2

Edwin Hubble

Edwin Powell Hubble was an American astronomer. Edwin Hubble was born on November 20, 1889 in Marshfield, Missouri. He studied astronomy in the University of Chicago. Hubble completed a doctorate in astronomy, enlisted in the U.S. Army and served a tour of duty in World War I.

While working at Mount Wilson, Hubble proved that other galaxies existed outside of the Milky Way, where Earth is located, by taking photos through the observatory's Hooker telescope and comparing the varying degrees of luminosity among Cepheid variable stars. In 1936, Hubble published *The Realm of the Nebulae*, his research in the field of extragalactic astronomy. Hubble worked at Mount Wilson Observatory until 1942, when he left to work at the Aberdeen Proving Grounds in Maryland during World War II. For his service during the war, in 1946, Hubble received the Medal of Merit. He died on September 28, 1953 for cerebral thrombosis. Hubble's work in the field of astronomy was truly revolutionary. He is also the recipient of the Franklin Medal (in physics), Legion of Merit, Bruce Medal and Gold Medal (from the Royal Astronomical Society). In 1990, 101 years after Hubble's birth, NASA launched the Hubble Space Telescope into orbit around Earth. The telescope, a tribute to him, has provided several of information about the cosmos, transmitting hundreds of thousands of images.

The distant galaxies in all directions are moving away from the earth, as seen by redshifts. Hubble's law describes this expansion. But one thing we have to remember that other celestial objects moving away from us does not imply that we are the center of the Universe. No matter where you are located in the Universe, you would see the same phenomenon happening in the same manner. So, there is no center of the Universe. All objects in the Universe are moving away from each other with the expansion of the Universe. This can be understood by a simple balloon analogy. At first, cover a balloon in dots with a marker and try to keep them equidistant. These dots represent galaxies or other astronomical objects. Now inflate the balloon. We will see that as the balloon increase in size. We will see that as the balloon increase in size, the dots will appear to be moving away from each other (see figure 10.3). But the dots themselves are not moving, the surface of the balloon is causing them to move apart. Here the surface of the balloon represents space-time. So, from here we can understand that the galaxies do not move away from each other through space themselves, its space itself that is expanding and carrying the galaxies with it.

Figure 10.3

Now we see that how we can get Hubble's constant from the redshift observations of distant galaxies. We know that the wavelengths of the emission and absorption lines from elements of the galaxies are redshifted due to their recessional velocity. If we can measure how much they are redshifted, we can calculate their velocity with the given formula below

$$v = C \left[\frac{\lambda - \lambda_0}{\lambda_0} \right]$$

Here λ is the wavelength of a line in the galaxy, λ_0 is the wavelength of the line in a lab on earth, v is the velocity of the galaxy and C is the speed of light.

As for example, from the relative intensity vs wavelength graph of the galaxy NGC 1357 we get the wavelengths of Ca K and Ca H in the lab frame and moving star frame.

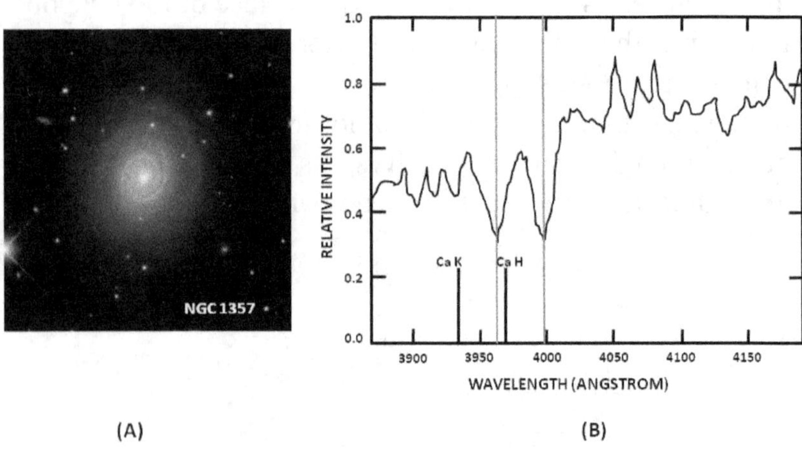

Figure 10.4: (A) NGC 1357 galaxy

(B) The graph of relative intensity of the spectrum from the moving galaxy and the wavelength of the absorption lines. Deep bold lines are the wavelength of Ca K and Ca H in the lab. Fade black lines represents observed wavelengths.

Now in the lab the wavelength of Ca K is 3933.7 Å but from observation, we get the wavelength as 3962 Å.
So by the formula, we get,

$$v = 3 \times 10^5 \left[\frac{3962 - 3933.7}{3933.7}\right] Kms^{-1} = 2158\ Kms^{-1}$$

Similarly in the lab the wavelength of Ca H is 3968.5 Å but from observation, we get the wavelength as 3998 Å.
So by the formula, we get,

$$v = 3 \times 10^5 \left[\frac{3998 - 3968.5}{3968.5}\right] Kms^{-1} = 2230\ Kms^{-1}$$

So the velocity is the average of these two i.e.

$$\left[\frac{2158 + 2230}{2}\right] Kms^{-1} = 2194\ Kms^{-1}$$

Now if we plot the Distance in Mpc *vs* recessional velocity in Kms^{-1} for some galaxies then we can get a graph. The slope of the mean line in this graph or the gradient is the value of Hubble's constant H_0 (see figure 10.1). This is approximately $72\ Kms^{-1}/Mpc$.

GALAXY	DISTANCE (Mpc)	RECESSIONAL VELOCITY (Km/Sec)
NGC-5357	0.45	200
NGC-3627	0.9	650
NGC-4472	2	850
NGC-1357	26	2200
NGC-1832	31	2000
NGC-7469	65	4470
NGC-5548	67	5270

AGE OF THE UNIVERSE FROM HUBBLE'S CONSTANT

We also can get the age of our Universe from this Hubble's constant with some simple concept and mathematics.
We know,

$$1\,pc = 3.09 \times 10^{16} m$$

$$1\,Mpc = 3.09 \times 10^{16} \times 10^{6} m = 3.09 \times 10^{22} m$$
$$= 3.09 \times 10^{19} Km$$

$$H_0 = \frac{72}{3.09 \times 10^{19}} \left(\frac{Kms^{-1}}{Km}\right) = 2.33 \times 10^{-18} s^{-1}$$

Now,

$$t = \frac{1}{H_0} = \frac{1}{2.33 \times 10^{-18}} = 4.29 \times 10^{17}\,s = 1.36 \times 10^{10}\,years$$
$$= 13.6\,billion\,years$$

This value is very near to the current measurement of the age of the Universe, which is 13.787 ± 0.020 billion years.

CHAPTER 11
COSMOLOGICAL CONSTANT AND DARK ENERGY

"The missing link in cosmology is the nature of dark matter and dark energy."
Stephen Hawking

Obviously, the term 'Dark energy' sounds more like an evil entity or related to black magic to whom who listen to this term for the first time. But, actually, it is not like that. Dark energy is the most vital invisible part of our universe which covers almost 70% of the whole universe. It is an unknown form of energy and we still don't know what it is like or why it exists. Dark energy is even more mysterious than Dark matter.

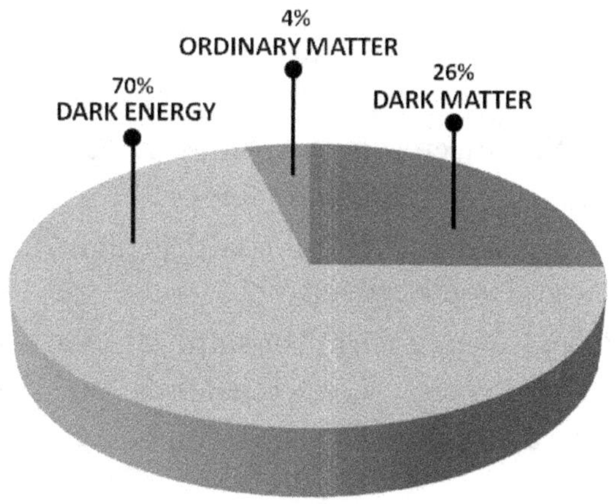

But, before we discuss this, I want to go back to 1915 again. Why this year is so important? Because we all know that in that year Albert Einstein gave us his famous general theory of relativity. From this theory, he gave 16 partial differential equations that relate the curvature or geometry of space-time with a mass within it. These are called 'Einstein's field equations' or simply 'Einstein's equations'. Actually, Einstein published these equations in the form of tensor equations. Tensor is one of the most complicated and difficult concepts of mathematics. But, here you no need to go deep into that. The first formulated version of Einstein's equations is given below

$$R_{\mu v} - \frac{1}{2} \mathcal{R} g_{\mu v} = \frac{8 \pi G}{C^4} T_{\mu v}$$

When Albert Einstein was developing his theory, astronomers of that era thought that the universe was static. But there was a problem. According to Einstein's this equation predicted that gravity pulled all matter into a point. Then Einstein thought that his theory may be wrong. So to fix his theory, he added a term to this equation, which is $\Lambda g_{\mu v}$. Thus the equation became,

$$R_{\mu v} - \frac{1}{2} \mathcal{R} g_{\mu v} + \Lambda g_{\mu v} = \frac{8 \pi G}{C^4} T_{\mu v}$$

Here $R_{\mu v}$ is the Ricci curvature tensor, $\mathcal{R}$ is the scalar curvature, $g_{\mu v}$ is the metric tensor, G is Newton's gravitational constant, C is the speed of light in vacuum, and $T_{\mu v}$ is the stress-energy tensor.

This Λ is called 'Cosmological constant'. It is to oppose the gravitational collapse. Now Einstein's field equation supports the astronomical observations. But, in 1930

Edwin Hubble showed that galaxies are moving away from us and hence the universe is not static, it's expanding. Then the term Λ was no longer needed. Einstein later considered the inclusion of that term was the biggest blunder of his life, according to his fellow physicist George Gamow. But, that was not the end of the story of Λ. Even the mistake of Albert Einstein had a great impact on the later studies of astronomy. In the 1990s, one thing was very mysterious about the expansion of the universe. The expansion of the universe should be stopped or it should be recollapsed if it has enough energy density. Maybe the universe will never stop expanding if it has little energy density. But, the expansion must be slow because the universe is full of matter and the attractive force of gravity pulls them together. In 1998, Hubble Space Telescope showed that, a long time ago, the universe was actually expanding more slowly than it is today. So the expansion is accelerating.

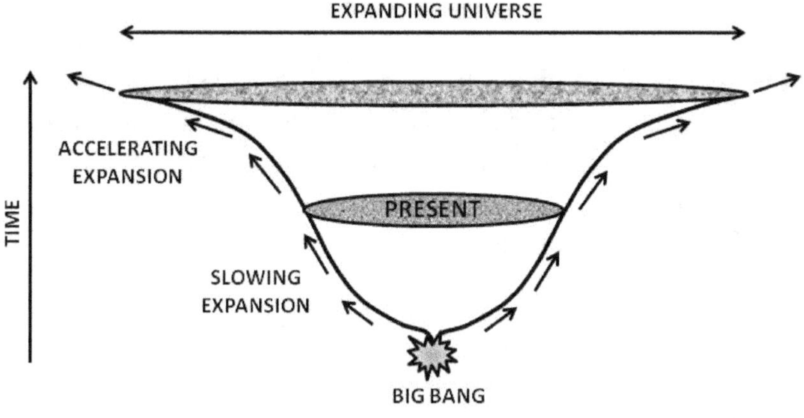

Figure 11.1

It was strange! No one knew how to explain it. There should be something that counteracts against gravity. Then they noticed that Einstein's cosmological constant Λ may explain it. Λ was back into existence again. Positive values of this constant suggest an exponential expansion of the universe.

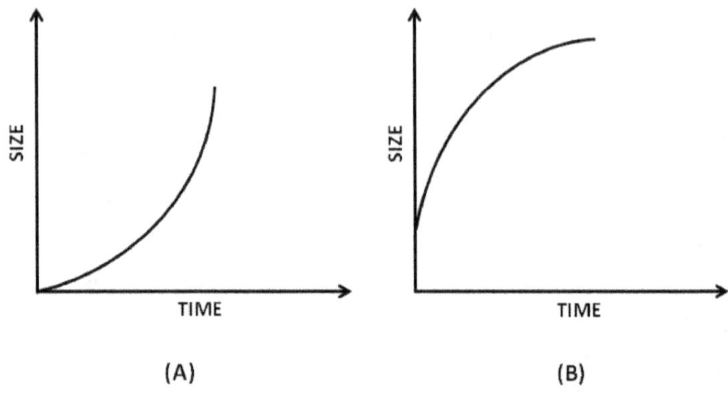

Figure 11.2: (A) Exponential expansion in the Universe's size vs. Time plot because of Λ (B) Graph of Universe's size vs. Time without Λ term in the field equation

Alexander Alexandrovich Friedmann solved Einstein's equations in 1922 and derived two independent equations called Friedmann equations. He was the first person to mathematically predict an expanding Universe.

The first Friedmann equation is

$$\frac{\dot{a}^2 + kC^2}{a^2} = \frac{8\pi G\rho + \Lambda C^2}{3}$$

The second Friedmann equation is

$$\frac{\ddot{a}}{a} = -\frac{4\pi G}{3}\left(\rho + \frac{3p}{C^2}\right) + \frac{\Lambda C^2}{3}$$

a is the scale factor and $\frac{\ddot{a}}{a}$ = H is the Hubble parameter. For the cosmological constant (Λ), we have a positive acceleration of the scale factor $(\ddot{a})$. This means that the expansion rate of the universe is increasing. But, without Λ the $(\ddot{a})$ will be negative because the density of the universe (ρ) and pressure (p) both slow down the expansion of the universe.

Physicists now think that the acceleration of the expansion is driven by a repulsive force and this is analogous to the cosmological constant. For lack of a better name, they call this mysterious force 'Dark energy'. But even today, physicists have no clear explanation for dark energy. This force seems to be growing stronger as the universe expands because this force or energy is the property of space itself. When more space comes, more of this energy would appear and it would not be diluted. Another explanation of this energy comes from the quantum theory of matter. According to this so-called 'empty space' are not actually empty, it is full of virtual particles that continually form and then disappear. But when scientists calculated the energy, it didn't give a satisfactory result. Maybe there is some strange energy fluid that fills the space. Some people called it 'quintessence' after the fifth element of the Greek philosophers.

Our universe may be dark energy dominated today but in the past, the universe was smaller and denser and the matter densities were much higher (here matter means both normal matter and dark matter). About 9 billion years ago, the dark energy density was negligibly small. Dark matter may have arisen at the very beginning. Dark energy also thought to have always been here but become detectable when the universe was already billions of years old. With time, the density of matter goes down but the density of dark energy remains constant. The main problem is that we still don't know what it is like, what it interacts with or more importantly why and how it exists. As there are no particles of dark energy, there is no way to detect it directly. The great challenge of today's astronomy is to uncover the nature of this invisible component of the Universe because the fate of our universe depends on it. If we understand its properties then we can reveal more mysteries of our Universe.

Alexander Alexandrovich Friedmann

Alexander Friedmann was a Russian mathematician and physicist was born on 16 June 1888 in Saint Petersburg, Russia. He started his scientific career in 1913 at Pavlovsk Observatory in St. Petersburg. He is best known for his 1922 paper, "*On the Curvature of Space*", on the expanding universe published by the journal Zeitschrift für Physik on June 29, 1922. His solutions to Einstein's field equations provided early evidence of an expanding universe. The expansion of the universe was finally corroborated several years later by Edwin Hubble's observations in 1929. He made different cosmological models are possible depending on whether the curvature is zero, negative, or positive. Such models are called 'Friedmann universes'. At the young age of 37, he died from typhoid fever.

CHAPTER 12
THE DESTINY

To know the future you must return to the past

I don't remember where I got this quote, but I must have read this somewhere. So, to know how the Universe could end, we will look into the very beginning of this Universe. There are many theories about the origin of the Universe. But the most widely accepted explanation is the Big Bang theory. According to this theory the universe as we know it started with a small singularity. The whole Universe was compacted into a small point (only a few millimetres wide) with infinite density and intense heat called a Singularity. About 13.7 billion years ago this singularity exploded (called 'Bang') and from this explosion matter, energy, space and time, these four things were created. Now the universe was started to evolve. For better understanding, we can divide this evolution into two eras, the Radiation era and the

Matter era. After the Big Bang, the universe was dominated by radiation. This radiation era is made of some smaller stages. We can call those as epochs (epoch means a particular period of time). The earliest is the Planck epoch (just after the Big Bang). At that time there were no matters, only energy and super force existed. Now, what is the super force? The four forces of nature together called super force. At that time these four forces unified as one single force. But later, gravity split away from the super force. The next epoch is called the grand unification epoch (about 10^{-43} seconds after the Big Bang). This epoch ended when strong force broke away from the three remaining unified forces of nature. Next was the inflationary epoch (about 10^{-36} seconds after the Big Bang) during which the universe rapidly expanded. In this epoch, particles are formed like quarks, electrons. Then came the electroweak epoch (about 10^{-32} seconds after the Big Bang), when the last two forces, electromagnetic and weak, finally split off. Next stage was the quark epoch (about 10^{-12} seconds after the Big Bang). At this stage, the universe was still too hot and dense. Then, in the hadron epoch (about 10^{-6} seconds after the Big Bang), the universe cooled down enough for quarks to bind together and form protons and neutrons. In the last stages of the radiation era, lepton epoch (about 1 second after the Big Bang) and nuclear epoch (about 100 seconds after the Big Bang), nuclei were formed by neutron and proton. By this, they also created the first chemical element in the universe, Helium. Matter formed by the elements. So, it was the beginning of the matter era. This era has three epochs. The first was the atomic epoch (about 50,000 years after the Big Bang). In this stage, the universe's temperature cooled down enough for electrons to attach to nuclei for the first time and the second element Hydrogen was created. Next is the galactic epoch (about 200 million years after the Big Bang) during which galaxies were formed by the clusters of atoms. The last stage is the stellar epoch (about 3 billion years after the Big Bang). In this epoch, stars and other celestial objects were formed and helped shape the universe as we know it today.

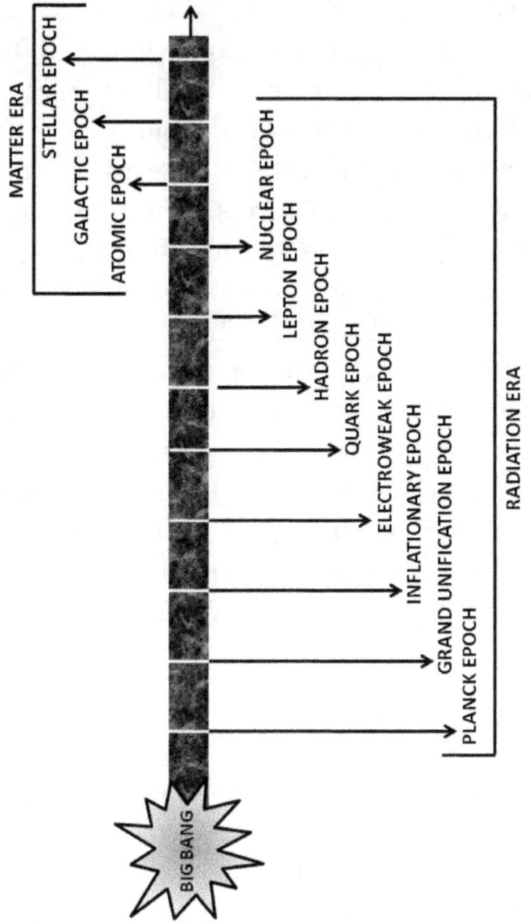

Figure 12.1: Radiation and matter era with different epochs

Now, let see, what happened with the dark energy. So for billions of years, the expansion began dominated by radiation, and then it cooled enough that matter became dominant. As the Universe expands continuously, the matter density has continued to drop while dark energy has remained constant. By the time the universe was 7.8 billion years old, the dark energy density reached about 33% of the total energy density of the Universe. It is an important value because it caused the expansion rate to begin accelerating. At present, about 13.8 billion years after the Big Bang, dark energy makes up about 70% of the total energy density of the Universe. Now we know that Dark energy has dominated our Universe since it was 7.8 billion years old and will determine the fate of our Universe.

The cosmological constant, density parameter, and dark energy, these three things are important to decide the ultimate fate of our Universe. I have discussed cosmological constant and dark energy in the previous chapter. Now, what is the density parameter? The density parameter Ω (omega) defined as the ratio of the average density of matter of the Universe and energy in the Universe to the critical density (the density at which the Universe would stop expanding). We can write Ω as

$$\Omega = \frac{\rho}{\rho_c}$$

Here ρ is the actual density of the Universe and ρ_c is the critical density.

It is the sum of a number of different components including normal and dark matter as well as dark energy. We can write it as

$$\Omega = \Omega_B + \Omega_D + \Omega_\Lambda$$

Here Ω_B is the density parameter for normal baryonic matter, Ω_D is the density parameter for dark matter and Ω_Λ is the density parameter for dark energy.

There are three different possible geometries depending on whether the value of Ω is equal to, less than or greater than 1. If Ω is less than 1, the Universe is open and will continue to expand forever. If Ω is greater than 1, the Universe is closed and the will recollapse because gravity will overcome the expansion. If Ω is exactly equal to 1, then the Universe is flat and contains enough matter to stop the expansion but not enough to recollapse it. There is also a fourth possibility, if $\Omega = 0$, then the Universe will be called an empty Universe with constant expansion rate. But, as it is nonphysical, we can ignore this one. All the above possibilities are illustrated below in figure 12.2.

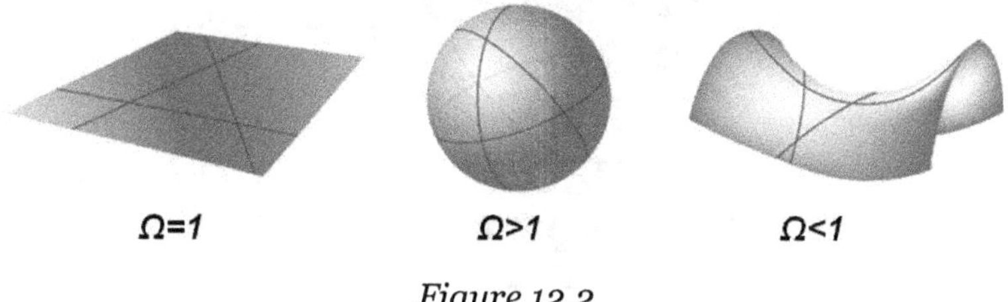

Figure 12.2

So the ultimate fate of the Universe depends on the cosmological constant, the shape of the Universe and most importantly, the change of dark energy density with the expansion of the universe. After knowing these factors, we can predict four scenarios. Those are Big Freeze, Big Rip, Big Crunch and Big Bounce.

When Ω less than 1, then the space-geometry is open and the Universe will be an open Universe. There are two possibilities, in this case, the Big Freeze or the Big Rip.

Imagine the expansion wouldn't be able to accelerate anymore, but the universe would keep getting bigger forever. The objects within the Universe - stars, planets, solar systems, clusters, galaxies would move away from each other floating separately in the vast space. The heat will disappear throughout space. That's why this scenario also called 'Heat Death'. The entropy will reach the maximum value and then heat in the universe will distribute evenly throughout the whole Universe. This means there will be no more scope for usable energy or heat and the Universe would die. In this case, no new stars can be born. The universe would become darker and colder, approaching a frozen state also called Big Chill.

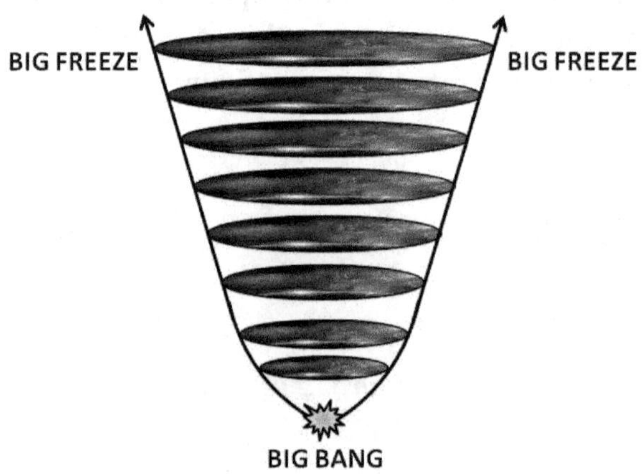

Figure 12.3: Big Freeze

Another version of Big Freeze is Big Rip (figure 12.4). In this scenario, the dark energy wins the battle against the gravitational and other binding forces (electromagnetic, weak, and strong nuclear forces that hold atoms and nuclei together). As a result, all objects of the Universe no matter how small, are ripped apart by the very strong repulsive force. They break into tiny pieces. Atoms and subatomic particles will be destroyed. At last, the energy density and expansion rate will become infinite.

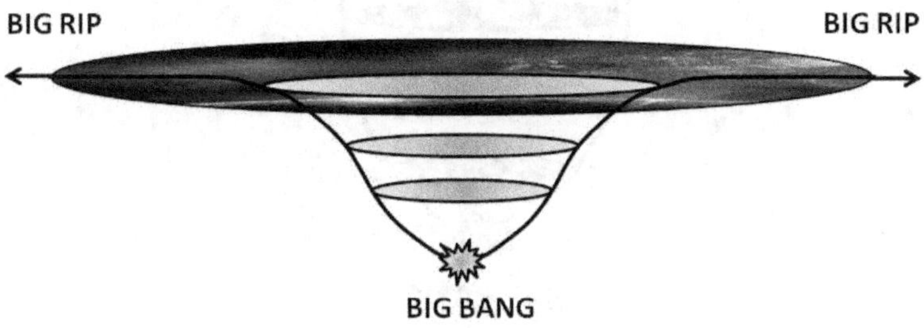

Figure 12.4: Big Rip

There is another scenario that is exactly opposite to the Big Bang. In this model, the Universe doesn't expand forever. At some point in the expansion, gravity will become the dominant force and it causes the Universe to shrink. All objects in the Universe will collide into each other and the Universe itself will collapse into a tiny point. It is like the reverse process of the Big Bang, well known as Big Crunch (figure 12.5). According to Big Bang theory, the Universe was created with an explosion of singularity, and according to Big Crunch, everything will implode back into a singularity. Another model tells us that the Big Bang can occur immediately after a Big Crunch of the previous Universe and it creates a new baby Universe. This process happened in a cyclic manner and we get an oscillatory model. This is also called the Big Bounce (figure 12.6).

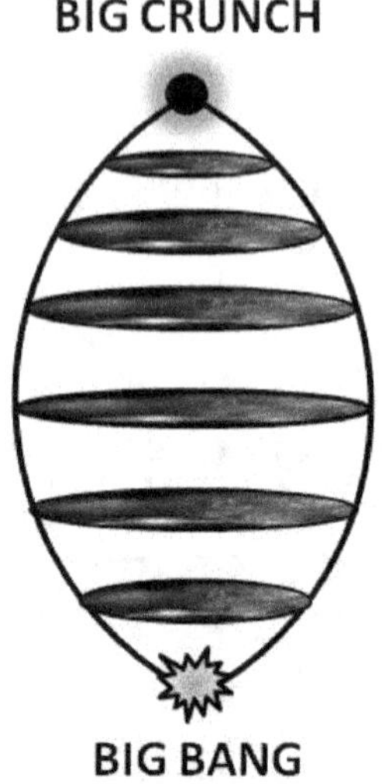

Figure 12.5: Big Crunch

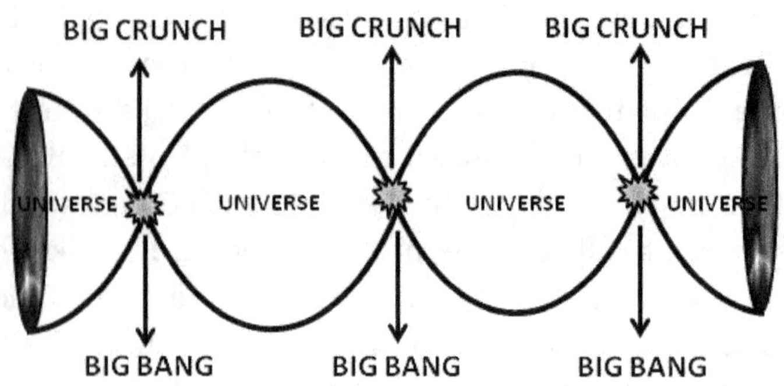

Figure 12.6: Big Bounce

So, what is the ultimate fate of the Universe among these models? According to the WMAP's (Wilkinson Microwave Anisotropy Probe)* measurement, the Big Freeze model is most favourable among these theories. But, the fact is that we have not a proper and deeper understanding of the dark energy and dark matter yet. There are much more unknown than known. So, the other three models still cannot be rejected.

Whenever we thought that we know everything and there is nothing left to be known, something new discovered that shook the whole world of science. As we are going into the deeper universe, we can see that we know almost nothing about it. There are so many things that we don't understand yet properly. But we should keep learning new things. Even though we might think that we know how everything is going to end, it's actually the beginning, because knowledge has no limit.

---

* WMAP is a NASA Explorer mission that launched in 2001 to make fundamental measurements of cosmology. Its mission ended in 2010. It measured temperature differences across the sky in the cosmic microwave background (CMB), the radiant heat remaining from the Big Bang. (For details read appendix 5, Page: 161-163)

APPENDIX

Appendix 1

Michelson-Morley Experiment

See carefully *Figure 2.4* and consider the following cases.
The path of beam 1
Time for beam 1 to travel from the beam splitter to mirror 1
and back is

$$t_1 = \frac{l_1}{c-v} + \frac{l_1}{c+v} = l_1\left(\frac{2c}{c^2 - v^2}\right) = \frac{2l_1}{c}\left(\frac{1}{1 - v^2/c^2}\right)$$

The speed of light is C and the speed of ether wind is v.
Upstream speed is $(C - v)$ and downstream speed is $(C + v)$ with respect to apparatus.
The path of beam 2

$$2\left[l_2^2 + \left(\frac{vt_2}{2}\right)^2\right]^{\frac{1}{2}} = ct_2$$

$$t_2 = \frac{2l_2}{\sqrt{c^2 - v^2}} = \frac{2l_2}{c}\frac{1}{1 - v^2/c^2}$$

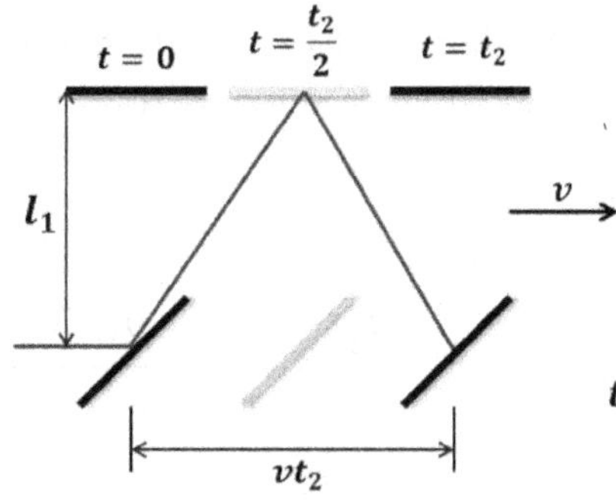

The difference in transit time is,

$$\Delta t = t_2 - t_1 = \frac{2}{c}\left[\frac{l_2}{\sqrt{1 - {v^2}/{c^2}}} - \frac{l_1}{1 - {v^2}/{c^2}}\right]$$

Now, if the instrument is rotated through 90°

$$\Delta t' = t'_2 - t'_1 = \frac{2}{c}\left[\frac{l_2}{1 - {v^2}/{c^2}} - \frac{l_1}{\sqrt{1 - {v^2}/{c^2}}}\right]$$

The rotation changes the difference by

$$\Delta t' - \Delta t = \frac{2}{c}\left[\frac{l_2 + l_1}{1 - {v^2}/{c^2}} - \frac{l_2 + l_1}{\sqrt{1 - {v^2}/{c^2}}}\right]$$

Using binomial expansion and ignoring terms higher than second order, we find,

$$\Delta t' - \Delta t \cong \frac{2}{c}(l_2 + l_1)\left[1 + {v^2}/{c^2} - 1 - \frac{1}{2}{v^2}/{c^2}\right] = \frac{(l_1 + l_2)}{c}{v^2}/{c^2}$$

So, the rotation should cause a shift in the fringe pattern due to the change of phase. Let, ΔN represents the number of fringes moving past the crosshairs as the pattern shifts. If the light of wavelength λ is used and the period of one vibration is T. Then,

$$\Delta N = \frac{\Delta t' - \Delta t}{T} \cong \frac{(l_1 + l_2)}{cT}{v^2}/{c^2} = \frac{(l_1 + l_2)}{\lambda}{v^2}/{c^2}$$

In Michelson-Morley experiment the arms were of nearly equal length, so, $l_1 = l_2 = l$

$$\Delta N = \frac{2l}{\lambda} v^2 / c^2$$

If we choose, $\lambda = 5.5 \times 10^{-7} m.$ $v/c = 10^{-4}$ then we obtain $\Delta N = 0.4$
This is a Shift of four-tenths a fringe!

Trials of this experiment done by various Scientists:

NAME	LOCATION	YEAR	FRING SHIFT MEASURED	NULL RESULT
Michelson	Potsdam	1881	≤ 0.02	$\approx$ YES
Michelson and Morley	Cleveland	1887	≤ 0.01	$\approx$ YES
Morley and Miller	Cleveland	1902-1904	≤ 0.015	YES
Miller	Mt. Wilson	1921	≤ 0.08	UNCLEAR
Miller	Cleveland	1923-1924	≤ 0.03	YES

Miller (Sunlight)	Cleveland	1924	≤ 0.014	YES
Tomaschek (Starlight)	Heidelberg	1924	≤ 0.02	YES
Miller	Mt. Wilson	1925-1926	≤ 0.088	UNCLEAR
Kennedy	Pasadena/Mt.Wilson	1926	≤ 0.002	YES
Illingworth	Pasadena	1927	≤ 0.0004	YES
Piccard and Stahel	With a balloon	1926	≤ 0.006	YES
Piccard and Stahel	Brussels	1927	≤ 0.0002	YES
Piccard and Stahel	Rigi	1927	≤ 0.0003	YES
Michelson et al.	Mt. Wilson	1929	≤ 0.01	YES
Joos	Jena	1930	≤ 0.002	YES

Appendix 2

Relativistic Doppler Effect Derivation

In chapter 9, I discussed the Relativistic Doppler effect. Here I derive the expression of the *Longituddional Doppler Effect.*

Consider two frames of references S and S'. S' is moving with a velocity v relative to S along the positive x-axis. Now, let the transmitter (source) and the receiver (observer) be situated at origins O and O' of frames S and S' respectively as shown in the figure.

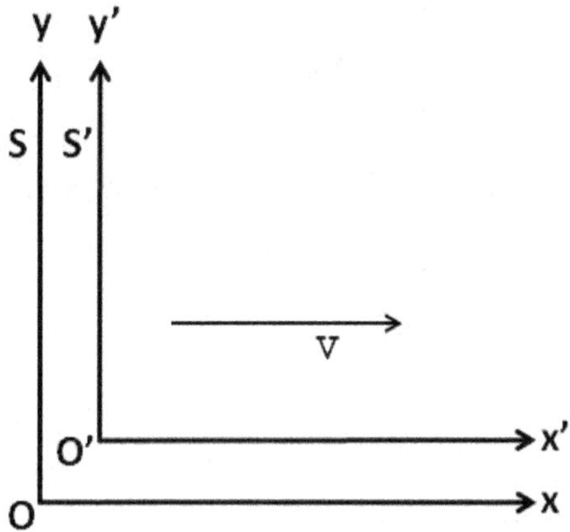

Let two light signals are transmitted at time $t = 0$ and $t = T$. T is the true period of light pulses. $\Delta t'$ be the interval between the reception of these pulses by the receiver at O'.

Here $\Delta t'$ is the proper time interval T' between these pulses.

We knew about Lorentz transformation in the chapter. Now from the inverse Lorentz transformation, we get,

$$x = \left(\frac{x' + vt'}{\sqrt{1 - v^2/c^2}} \right)$$

$$\Delta x = \left(\frac{\Delta x' + v\Delta t'}{\sqrt{1 - v^2/c^2}} \right)$$

$$= \left(\frac{v\Delta t'}{\sqrt{1 - v^2/c^2}} \right)$$

$$= \left(\frac{vT'}{\sqrt{1 - v^2/c^2}} \right) \quad \ldots\ldots\ldots\ldots \text{(1)}$$

$\Delta x' = 0$ Because the observer continues to be at O' all the time the distance Δx covered by him in frame S' during the reception of two pulses is zero.

Equation 1 shows that the second pulse has to travel his much distance Δx more than the first pulse in frame S along the x-axis to be able to reach at the origin O' in the moving frame S'.

Now from inverse Lorentz transformations of t we get,

$$t = \left\{ \frac{t' + \left(\frac{vx'}{c^2}\right)}{\sqrt{1 - v^2/c^2}} \right\}$$

$$\Delta t = \left\{ \frac{\Delta t' + \left(\frac{v\Delta x'}{c^2}\right)}{\sqrt{1 - v^2/c^2}} \right\} = \left(\frac{\Delta t'}{\sqrt{1 - v^2/c^2}} \right)$$

$$\Delta t = \left\{ \frac{T'}{\sqrt{1 - v^2/c^2}} \right\} \quad \ldots\ldots\ldots\ldots\ldots \quad (2)$$

Here $\Delta t = \left(T + \frac{\Delta x}{c}\right)$ $\quad \ldots\ldots\ldots\ldots \quad (3)$

T is the actual time period and $\frac{\Delta x}{c}$ is the time taken by the second pulse to cover the extra distance Δx in the frame S.

Substituting the values of Δx and Δt from equations 1, 2 and 3, we get,

$$\frac{T'}{\sqrt{1 - v^2/c^2}} = T + \frac{vT'}{c\sqrt{1 - v^2/c^2}}$$

From here by solving we get an equation 4 as,

$$T = \frac{T'\left(1 - \frac{v^2}{c^2}\right)}{\sqrt{1 - v^2/c^2}} = T'\sqrt{\left(\frac{1 - \frac{v}{c}}{1 + \frac{v}{c}}\right)}$$

If ϑ and ϑ' be the actual and observed frequencies of light pulses respectively.

Then, $\vartheta = \frac{1}{T}$ and $\vartheta' = \frac{1}{T'}$. Put these values in equation 4. Now we get,

$$\frac{1}{\vartheta} = \frac{1}{\vartheta'} \sqrt{\left(\frac{1 - \frac{v}{c}}{1 + \frac{v}{c}}\right)}$$

$$\vartheta' = \vartheta \sqrt{\frac{\left(1 - \frac{v}{c}\right)}{\left(1 + \frac{v}{c}\right)}}$$

We know that $\lambda = \frac{c}{\vartheta}$ and $\lambda' = \frac{c}{\vartheta'}$ where λ and λ' are actual and apparent wavelength respectively.

$$\frac{c}{\lambda'} = \frac{c}{\lambda} \sqrt{\frac{\left(1 - \frac{v}{c}\right)}{\left(1 + \frac{v}{c}\right)}}$$

In terms of wavelength we get,

$$\lambda' = \lambda \sqrt{\frac{\left(1 + \frac{v}{c}\right)}{\left(1 - \frac{v}{c}\right)}}$$

This is the expression of the *Longitudinal Doppler Effect.*

Appendix 3

Particle Physics in brief

Particle physics is a part of physics that studies the nature of the particles that constitute matter and radiation. The first elementary particle discovered was electron in 1898. Then the existence of positively charged protons inside the nucleus was indicated by the scattering experiment in 1914. With the discovery of the neutron in 1932, the total number of elementary particles became three. But, then the cosmic ray research and research on the interaction of fundamental particles caused a dramatic increase in the number of elementary particles. With the powerful accelerators and detectors new types of particles discovered, produced and detected in nuclear reactions. Many elementary particles are the product of high energy collisions. To understand the significance of these particles we have to know about fundamental interactions. All interactions are transmitted from one particle to others by successive processes. There are four types of fundamental interactions. These are Gravitational, Electromagnetic, Weak and strong interactions. Gravitational interaction is the weakest of all the fundamental interactions. It acts between all bodies having mass and is described by long-range inverse square type Newtonian law of gravitation. This gravitational interaction becomes inappreciable for the interaction of elementary particles, nuclei and atoms. Electromagnetic interaction acts between charged particles and particles with electric and magnetic moments. It is much stronger than the previous one and described by long-range inverse square type coulomb's law. The exchange particle in this interaction is 'photon' which is massless. The weak interaction is a short-range force. The range is smaller than 0.001 fm (femtometre) and mediated through bosons particles. The 'electroweak theory' unified electromagnetic and weak interactions in 1960. The strong interaction is the strongest force in nature and acts between nucleons in a nucleus (between a neutron and a proton). This is a very short range, charge-independent attractive force.

The elementary particles may be classified in a number of different ways depending on their masses, interactions, statistics etc. Commonly they have classified into three broad classes: Gauge Bosons, Leptons and Hadrons. Gauge bosons described by gauge theory; include photons, gluons, W bosons and Z bosons particles. A hypothetical particle graviton is also a boson.

Leptons are weakly interacting particles. There are now twelve leptons (six leptons with their six antiparticles). They are the electron, positron, pair of muons, pair of tauons, three neutrinos: electron-neutrino, muon-neutrino, tauon-neutrino and three antineutrinos. Leptons are appeared to have no internal structures. They are all fermions as they obey Fermi-Dirac statistics. Among them, only muon and tauon are unstable.

Hardons have strong nuclear interactions. They are sub-divided into mesons and baryons. Mesons are strongly interacting particles and include pions, kaons and etaons. Hardons of half-integer spins are called baryons. Proton, neutron, lambdaon, sigmaon, xions, omegaon and their antiparticles are all baryons. These are also obeying Fermi-Dirac statistics.

Many attempts were made to see if the elementary particles can be understood in terms of the simpler and basic units of particles. One of them is the Quark Model given by Gell-Mann and Zweig in 1963. They proposed that hadrons are made up of still smaller particles called quarks. They have fractional charges. All quarks have spin half and they are fermions. There are six types of quarks: Up (u), Down (d), Strange (s), Charm (c), Top (t) and Bottom (b). Baryons are composed of three quarks each but mesons are composed of two quarks each. For example, the quark content of proton is *uud*, that of the neutron is *udd* etc.

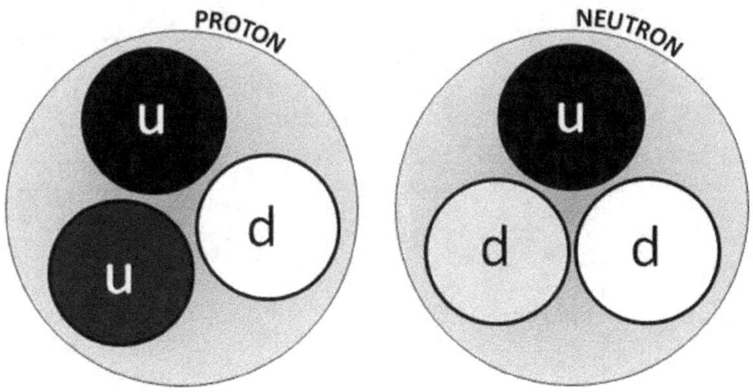

Each of six types of quarks (u,d,c,s,t,b) are given three distinct colours or flavours as red, blue and green. So similer quarks in a hardon are with a different flavour.

The strong interactions of quarks are assumed to be mediated by eight massless vector bosons, which are named 'Gluon'. Gluons act as glue and bind quarks together. We can not detect a free gluon as individual gluon cannot be isolated.

'Standard model' is a theory of fundamental particles and how they interact. This name was given in the 1970s. A standard model of sub-atomic particles has been proposed and most of the experimental information about them can be theoretically described with its help. There are seventeen particles in the standard model. These are: six quarks, six leptons, four gauge boson and recently added Higgs boson (in 2012). The standard model has much success but it also has some drawbacks. The next step is to combine all types of interactions into a single one, and the theory that attempts to do this is called the Grand Unified Theory (GUT). The GUT has wide implications in astrophysics.

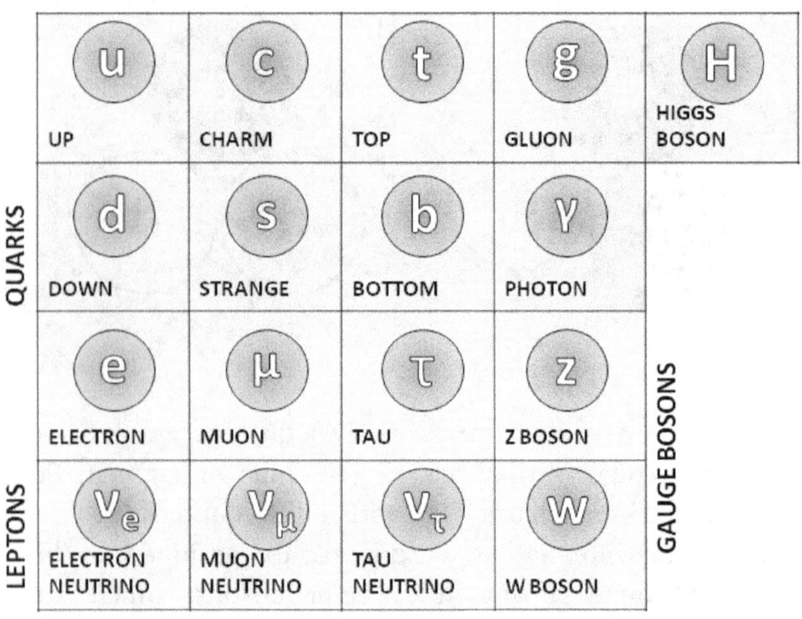

Standard model

Appendix 4

Euclidean and Non-Euclidean Geometry

To understand the theories of the Universe and space-time curvature in a better way, we must have an idea about the different types of geometry and about their uses. Euclidean geometry is a mathematical system named after Greek mathematician Euclid. It is described in his textbook of geometry, the Elements. This is the plane geometry, still taught in secondary school. Euclidean geometry is mainly a systematic and logical representation of earlier knowledge of geometry. We all know, according to this geometry, if a straight line and a point not on that line given then there is only one line you can draw that passes through that point and is parallel to the first line. It is Euclid's parallel postulate. So, you can see this geometry is flat and its curvature is zero. But the main problem is this geometry defines the situations of the plane only. When we are dealing with different surfaces, we are faced many drawbacks of solving problems through the same geometry. This geometry is also called parabolic geometry. If you try to apply the same geometry on the surface of a football then you will face problems. But, we always need those theorems which can be applied to every situation. In the previous chapters, we saw that our universe is not so flat, space and time can be curved. So, now we need some better geometry to explain the laws of the universe. Carl Friedrich Gauss and Ferdinand Karl Schweikart had the ideas of non-Euclidean geometry at the beginning of the 19th century. But, they did not publish their thought. Then, in 1830, Janos Bolyai and Nikolai Ivanovich Lobachevsky separately published treatises on hyperbolic geometry. It is also called Bolyai-Lobachevsky geometry. In this geometry, the surface is negatively curved. This leads to a variation of Euclid's parallel postulate. According to this, for a given straight line and a point not on the line there exist an infinite number of straight lines through the point parallel to the original line. Then in 1854, Bernhard Riemann founded a special type of geometry, called Riemannian geometry. It discusses the ideas of manifolds, Riemannian metric and Riemannian curvature. It is also called spherical or elliptical geometry. Here the surface is positively curved. In this geometry, Euclid's parallel postulate has the form like this: for a given straight line and the point not on the line, there are no straight lines through the point parallel to the original line. We know that the sum of angles of a triangle is 180 degrees in Euclidean geometry. We all study it from school. But, it is very interesting that for hyperbolic geometry, the sum of the angles of a triangle is less than 180 degrees and for Riemannian or elliptical geometry the sum of angles is always greater than 180 degrees. You can easily understand this concept by the given figures.

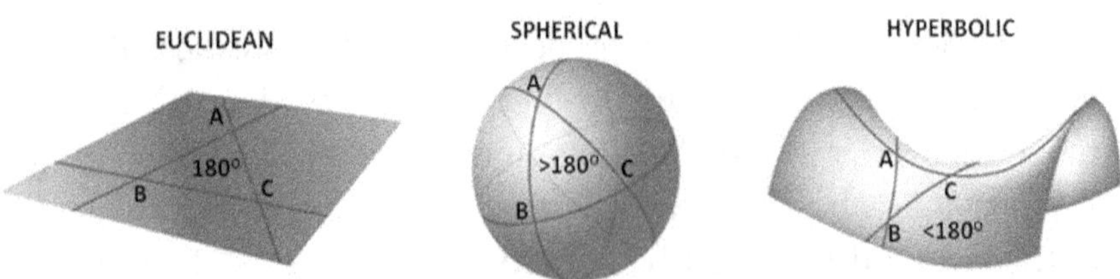

Appendix 5

WMAP

(Wilkinson Microwave Anisotropy Probe)

The Wilkinson Microwave Anisotropy Probe (WMAP) was a spacecraft operating from 2001 to 2010 which measured temperature differences across the sky in the cosmic microwave background (CMB). It has put fundamental theories of the nature of the universe to a precise test. This mission gave us the first best ever map of the baby Universe showing variations in the brightness of radiation from primordial matter. The MAP mission was proposed to NASA in 1995, coincidentally my birth year. It was selected for the definition study in 1996 and approved for development in 1997. WMAP launched on June 30, 2001 and the mission duration was 9 years 1 month 2 days (from launch to end collection of science data). It was initially called the Microwave Anisotropy Probe (MAP); the satellite was renamed in 2003 to honour the memory and accomplishments of David T. Wilkinson, a member of the science team and a pioneer in CMB studies. Its observing station was near the "second Lagrange point" of the Earth-Sun system, a million miles from Earth in the direction opposite the sun.

WMAP's "baby picture of the universe" maps the afterglow of the hot, young universe at a time when it was only

375,000 years old when it was a tiny fraction of its current age of 13.77 billion years. The first results were issued in February 2003, then with major updates in 2005, 2007, 2009, 2011, and on December 20, 2012, the nine-year WMAP data and related images were released. The study found that 95% of the early universe is composed of dark matter and dark energy, the curvature of space is less than 0.4% of "flat" and the universe emerged from the cosmic Dark ages 'about 400 million years' after the Big Bang. One possibility for the dark energy, Albert Einstein's "cosmological constant," is fully consistent with the WMAP data. WMAP has also provided the timing of epoch when the first stars began to shine when the universe was about 400 million old. The main result of the mission is contained in the various oval maps of the CMB temperature differences. These oval images present the temperature distribution derived by the WMAP team from the observations by the telescope during the mission. The pattern of these variations fits the predictions of a physical theory called inflation, which suggests that during the first split second of existence the universe expanded incredibly fast.

'Planck' is a follow-on mission of this mission. On 21 March 2013, the European-led research team behind the Planck cosmology probe released the mission's all-sky map of the cosmic microwave background. Based on the 2013 data, the universe contains 4.9% ordinary matter, 26.8% dark matter and 68.3% dark energy. On 5 February 2015, new data was released by the Planck mission, according to which the age of the universe is 13.799 ± 0.021 billion years old and the Hubble constant was measured to be 67.74 ± 0.46 (km/s)/Mpc.

The full-sky image of the temperature fluctuations (shown as colour differences) in the cosmic microwave background, made from nine years of WMAP observations. These are the seeds of galaxies, from a time when the universe was under 400,000 years old.

Credits: NASA/WMAP Science Team

Appendix 6

SYMBOLS USED IN THIS BOOK

v	Velocity
l	Length
M, m	Mass
R	Radius
$R_{\odot}$	Radius of the Sun
$M_{\odot}$	Mass of the Sun
R_s	Schwarzschild radius
t	Time
E	Energy
T	Temperature
h	Plank's constant (approximately 6.626176×10^{-34} joule-seconds.in SI Unit)
G	Gravitational constant
ϑ, f	Frequency
λ	Wavelength
D	Distance

π	*Pi (approximately equal to 3.14159)*
Λ	*Cosmological constant*
H_0	*Hubble's constant*
$R_{\mu\nu}$	*Ricci curvature tensor*
$g_{\mu\nu}$	*Metric tensor*
$T_{\mu\nu}$	*Stress-Energy tensor*
a	*Scale factor*
p	*Pressure*
ρ	*Density*
ρ_c	*Critical density*
Ω	*Density parameter*
$\mathcal{R}$	*Scalar curvature*
K	*Boltzmann constant (approximately 1.380649×10^{-23} J·K^{-1} in SI Unit)*

BIBLIOGRAPHY

Books and articles:

1. Einstein, A. (2017). *Relativity*. Fingerprint Publishing.

2. Hawking, S. (1992). *Stephen Hawking's A brief history of time* (pp. 85-105). New York: Bantam Books.

3. Hawking, S. (2001). *The universe in a nutshell*. New York: Bantam.

4. Narlikar, J. (2005). *Black holes*. New Delhi: National Book Trust, India.

5. Resnick, R. *Introduction to special relativity*. Wiley india Pvt. Ltd.

6. Piccioni, R. (2010). *Einstein for Everyone* (1st ed.). Jaico.

7. Hawking, S. (2016). *Black holes*. London: Bantam Books.

8. Griffiths, D. (2014). Introduction to elementary particles (2nd ed.). Weinheim: Wiley-VCH Verlag.

9. Accelerating Universe and Dark Energy - The Big Bang and the Big Crunch - The Physics of the Universe. (2020). https://www.physicsoftheuniverse.com/topics_bigbang_accelerating.html

10. MACHOs | COSMOS. (2020). http://astronomy.swin.edu.au/cosmos/M/MACHOs

11. Hawking, S. (1999). A brief history of relativity - December 31, 1999. http://edition.cnn.com/ALLPOLITICS/time/1999/12/27/relativity.html

12. Choi, C. (2020). Our Expanding Universe: Age, History & Other Facts. https://www.space.com/52-the-expanding-universe-from-the-big-bang-to-today.html

13. Ranzan, C. (2020). Models of the Universe. http://www.cellularuniverse.org/UniverseModels.htm#SS1

14. Pivarski, J. (2012). *Dark Matter and Dark Energy* [EBook]. http://coffeeshopphysics.com/talks/2012-03-04_Fermilab_ask-a-scientist.pdf

15. Hawking, S. (2008). Into a Black Hole. http://www.hawking.org.uk/into-a-black-hole.html

16. Gravity. (2020). https://en.wikipedia.org/wiki/Gravity

17. Hubble site (2020). http://hubblesite.org/

18. LIGO | MIT. (2020). https://space.mit.edu/LIGO/

19. Hawking, S. (1996). The Beginning of Time. http://www.hawking.org.uk/the-beginning-of-time.html

20. Wilkinson Microwave Anisotropy Probe (WMAP). https://map.gsfc.nasa.gov/

21. Cosmological constant. (2020). https://en.wikipedia.org/wiki/Cosmological_constant

22. Event Horizon Telescope. https://eventhorizontelescope.org/

23. WMAP 9 Year Mission Results. https://wmap.gsfc.nasa.gov/news

24. Wilkinson Microwave Anisotropy Probe. https://en.wikipedia.org/wiki/Wilkinson_Microwave_Anisotropy_Probe

25. Wikipedia

Collection of photographs:

1. Case Western Reserve University/ Wikimedia Commons

2. Space Telescope Science Institute (STScI)

3. National Aeronautics and Space Administration (NASA) Gallery

4. Hubble Site

5. Double Negative (DNEG)

6. NASA, ESA, Martin Kornmesser (ESA/Hubble)

7. NRAO/AUI/NSF

8. EHT collaboration / Astrophysical journal letters, 875(2019) L1/CC BY 3.0

9. EHT Collaboration

10. NASA/WMAP Science Team

11. ESA/Hubble

12. Figure 1.1, 1.2, 1.3, 2.1, 2.3, 2.4, 2.8, 2.9, 2.10, 3.1, 3.2, 3.3, 3.4, 4.5, 5.1, 5.2, 5.3, 5.5, 5.6, 5.8, 5.9, 5.10, 6.2, 6.4, 7.1, 7.2, 7.3, 8.1, 8.2, 9.1, 9.2, 9.3, 9.4, 9.5, 10.1, 10.2, 10.3, 10.4, 11.1, 11.2, 12.1, 12.2, 12.3, 12.5, 12.6 and photographs of appendix part are designed by the author himself.

INDEX

Another book by the author

PHYSICS My Love
Story of Physics for Everyone

This is an elementary introduction to the fascinating world of Physics. The chief aim of this book is to increase the interest of school students and others about Physics. The subject matter is presented in a very simple way without mathematical calculations, so that, everyone can understand easily.

Available at
Amazon • Flipkart • Amazon Kindle
Notionpress • Google-Books
(Paperback and Kindle edition)

For more details visit
Official author website of Shuvadip Ganguli
Scan this QR Code